Lecture Notes in Physics

Volume 893

The Lecture Notes in Physics

The series Lecture Notes in Physics (LNP), founded in 1969, reports new developments in physics research and teaching-quickly and informally, but with a high quality and the explicit aim to summarize and communicate current knowledge in an accessible way. Books published in this series are conceived as bridging material between advanced graduate textbooks and the forefront of research and to serve three purposes:

- to be a compact and modern up-to-date source of reference on a well-defined topic
- to serve as an accessible introduction to the field to postgraduate students and nonspecialist researchers from related areas
- to be a source of advanced teaching material for specialized seminars, courses and schools

Both monographs and multi-author volumes will be considered for publication. Edited volumes should, however, consist of a very limited number of contributions only. Proceedings will not be considered for LNP.

Volumes published in LNP are disseminated both in print and in electronic formats, the electronic archive being available at springerlink.com. The series content is indexed, abstracted and referenced by many abstracting and information services, bibliographic networks, subscription agencies, library networks, and consortia.

Proposals should be sent to a member of the Editorial Board, or directly to the managing editor at Springer:

Christian Caron
Springer Heidelberg
Physics Editorial Department I
Tiergartenstrasse 17
69121 Heidelberg/Germany
christian.caron@springer.com

More information about this series at
http://www.springer.com/series/5304

François David

The Formalisms of Quantum Mechanics

An Introduction

 Springer

François David
Institut de Physique Théorique
CEA Saclay
Gif-sur-Yvette
France

ISSN 0075-8450   ISSN 1616-6361 (electronic)
Lecture Notes in Physics
ISBN 978-3-319-10538-3   ISBN 978-3-319-10539-0 (eBook)
DOI 10.1007/978-3-319-10539-0
Springer Cham Heidelberg New York Dordrecht London

Library of Congress Control Number: 2014954643

Printed on acid-free paper

Springer is part of Springer Science+Business Media (www.springer.com)

Contents

Chapter 1
Introduction

1.1 Motivation

Quantum mechanics in its modern form is now more than 80 years old. It is probably the most successful and complete physical theory that was ever proposed in the whole history of science. The quantum theory is born with the twentieth century from the attempts to understand the structure and the properties of matter, of light and of their interactions at the atomic scale. In a quarter of a century a consistent formulation of the quantum theory was achieved, and it became the general conceptual framework to understand and formulate the laws of the physical world. This theory is quantum mechanics. It is valid from the microscopic distances (the presently accessible high energy scales $10\,\mathrm{TeV} \simeq 10^{-19}\,\mathrm{m}$, and possibly from the Planck scale ($10^{-35}\,\mathrm{m}$), up to the macroscopic distances (from $\ell \sim 1\,\mathrm{nm}$ up to $\ell \sim 10^5\,\mathrm{m}$ depending of physical systems, from molecules to neutron stars, and experiments). Beyond these scales, for most scientists the "old" classical mechanics takes over as an effective theory, valid when quantum interferences and non-local correlations effects can be neglected.

Quantum mechanics has fully revolutionized and unified the fields of physics (as a whole, from particle and nuclear physics to atomic an molecular physics, optics, condensed matter physics and material science), and of chemistry (again as a whole). It had a fundamental impact in astronomy, cosmology and astrophysics, in biology. It had and still has a big influence on mathematics (with feedback). It had of course a huge impact on modern technology, communications, computers, energy, sensors, weaponry (unfortunately) etc. In all these fields of science, and despite the impressive experimental and technical progresses of the last decades quantum mechanics has never been found at fault or challenged by experiments, and its theoretical foundations are considered as very solid. These tremendous successes have overcome the doubts and the discussions about the foundational principles and the paradoxical and non-classical features of quantum mechanics and the possible interpretations of the formalism, that take place since its beginning.

© Springer International Publishing Switzerland 2015
F. David, *The Formalisms of Quantum Mechanics*, Lecture Notes in Physics 893,
DOI 10.1007/978-3-319-10539-0_1

Quantum information became a important and very active field (both theoreti-
cally and experimentally) in the last decades. It has enriched our points of view
on the quantum theory, and on its applications, such as cryptography and quantum
computing. Quantum information science, together with the experimental tests of
quantum mechanics on microscopic quantum systems, the theoretical advances in
quantum gravity and cosmology, the slow diffusion of the concepts of quantum
theory in the general public, etc. have led to a revival of the discussions about the
principles of quantum mechanics and its seemingly paradoxical aspects.

This development of the reflexions and of the discussions about quantum theory,
and the concurrent expansion of its successes and of its applications (possibly
together with the rise of sensationalism in science outreach and in the practice
of science. . .) give sometimes the feeling that quantum mechanics is both: (1)
the unchallenged and dominant paradigm of modern physical sciences, (2) but at
the same time a still mysterious and poorly understood theory, that awaits some
(imminent) revolution.

The purpose of these lecture notes is to discuss more the first point: why is
quantum mechanics so consistent and successful? They will present a brief and
introductory (but hopefully coherent) view of the main formalizations of quantum
mechanics (and of its version compatible with special relativity, namely quantum
field theory), of their interrelations and of their theoretical foundations.

Two formulations are the standard tools used in most applications of quantum theory in physics and chemistry. These are: (1) the "standard" formulation
of quantum mechanics (involving the Hilbert space of pure states, self-adjoint
operators as physical observables, and the probabilistic interpretation given by
the Born rule); and (2) the path integral and functional integral representations of
probabilities amplitudes. It is important to be aware that there are other formulations
of quantum mechanics, i.e. other representations (in the mathematical sense) of
quantum mechanics, which allow a better comprehension and justification of the
quantum theory. This course will focus on two of them: (1) the algebraic quantum
theory and quantum field theory, and (2) the so called "quantum logic" approach.
These are the formulations that I find the most interesting (besides the standard one)
and that I think I managed to understand (somehow. . .).

In my opinion discussing and comparing these various formulations is useful in
order to get a better understanding of the coherence and the strength of the quantum
formalism. This is important when discussing which features of quantum mechanics
are basic principles and which ones are just natural consequences of the former.
Indeed what are the principles and what are the physical consequences depends
on the precise formulations chosen. For instance, as we shall see, the Born rule or
the projection postulate are indeed postulates in the standard formulation. In some
other formulations, such as quantum logic, they are mere consequences of other
postulates.

There are many excellent books that present and discuss the principles of
quantum mechanics. Nowadays most of them treat to some extend the foundational
and interpretational issues. Nevertheless, usually only the standard formalism is
presented in some details, with its physical applications, and it is in this framework

that the questions of the principles and of the possible interpretations of the formalism are discussed: Bell-like inequalities, non-locality, determinism and chance, hidden variable models, Copenhagen interpretation versus the rest of the world, information-theoretic formulations, etc. Even in the excellent and highly advisable reviews by Auletta [8] and by Laloë [71] the algebraic formalism and quantum logic are discussed in a few pages. I think that discussing in more details these other formulations is quite useful and sheds lights on the meaning of the formalism and on its possible interpretations. It is also useful when considering the relations between quantum theory, information theory and quantum gravity. Thus I hope these notes will not be "just another review on quantum theory" (and possibly rather amateurish), but that they may fill some gap and be useful, at least to some readers.

1.2 Organization

After this introductory section, the second section of these notes is a reminder of the basic concepts of classical physics, of probabilities and of the standard (canonical) and path integral formulations of quantum physics. I tried to introduce in a consistent way the important classical concepts of states, observables and probabilities, which are of course crucial in the formulations of quantum mechanics. I discuss in particular the concept of probabilities in the quantum world and the issue of reversibility in quantum mechanics in the last subsection.

The third section is devoted to a presentation and a discussion of the algebraic formulation of quantum mechanics and of quantum field theory, based on operator algebras. Several aspects of the discussion are original. Firstly I justify the appearance of abstract C^*-algebras of observables using arguments based on causality and reversibility. In particular the existence of a $*$-involution (corresponding to conjugation) is argued to follow from the assumption of reversibility for the quantum probabilities. Secondly, the formulation is based on real algebras, not complex algebras as is usually done, and I explain why this is more natural. I give the mathematical references which justify that the GNS theorem, which ensures that complex abstract C^*-algebras are always representable as algebras of operators on a Hilbert space, is also valid for real algebras. The standard physical arguments for the use of complex algebras are only given after the general construction. The rest of the presentation is shorter and quite standard.

The fourth section is devoted to one of the formulations of the so-called quantum logic formalism. This formalism is much less popular outside the community interested in the foundational basis of quantum mechanics, and in mathematics, but deserves to be better known. Indeed, it provides a convincing justification of the algebraic structure of quantum mechanics, which for an important part is still postulated in the algebraic formalism. Again, if the global content is not original, I try to present the quantum logic formalism in a similar light than the algebraic formalism, pointing out which aspects are linked to causality, which ones to reversibility, and which ones to locality and separability. This way to present the quantum logic formalism is original, I think. Finally, I discuss in much more details than is usually done Gleason's theorem, a very important theorem of Hilbert space geometry and operator algebras, which justify the Born rule and is also very important when discussing the status of hidden variable theories.

The final section contains short, introductory and more standard discussions of some other questions about the quantum formalism. I present some recent approaches based on quantum information. I discuss some features of quantum correlations: entanglement, entropic inequalities, the Tisrelson bound for bipartite systems. The problems with hidden variables, contextuality, non-locality, are reviewed. Some very basic features of quantum measurements are recalled. Then I stress the difference between the various formalizations (representations) of quantum mechanics, the various possible interpretations of this formalism, and the alternative proposal for a quantum theory that are not equivalent to quantum mechanics. I finish this section with a few very standard remarks on the problem of quantum gravity.

1.3   What This Course is Not!

These notes are tentatively aimed at a non specialized but educated audience: graduate students and more advanced researchers in physics. The mathematical formalism is the main subject of the course, but it will be presented and discussed at a not too abstract, rigorous or advanced level. Let me stress that **these notes do not intend to be**:

– a real course of mathematics or of mathematical physics;
– a real physics course on high energy quantum physics, on atomic physics and quantum optics, of quantum condensed matter, discussing the physics of specific systems and their applications;
– a course on what is *not* quantum mechanics;
– a course on the history of quantum physics;
– a course on the present sociology of quantum physics;
– a course on the philosophical and epistemological aspects of quantum physics.

But I hope that it could be useful as an introduction to these topics. Please keep in mind that this is not a course made by a specialist, it is rather a course made by an amateur, but theoretical physicist, for amateurs!

Acknowledgements These lecture notes started from: (1) a spin-off of yet-to-be-published lecture notes for a master course in statistical field theory that I give at Ecole Normale Supérieure in Paris since 2001, and of courses in quantum field theory that I gave in the recent years at the Perimeter Institute, (2) a growing interest in quantum information science and the foundational discussions about quantum formalism (a standard syndrome for the physicist over 50... but motivated also, more constructively I hope, by my work during 5 years as an evaluator for the European Research Council), (3) a course for the Graduate School of Physics of the Paris Area (ED107) that I gave in

2012 at the Institut de Physique Théorique. A first version of the notes of this course is available online at [33].

Most of the content of these lecture notes is review material (I hope properly assimilated). A few sections and aspects of the presentation contain some original material. In particular the emphasis on the role of reversibility in the formalism (especially in Sects. 3 and 4) has been published in a short form in [32]. These notes can be considered partly as a very extended version of this letter.

I would like very much to thank for their interest, advices, critics and support at various stage of this work Roger Balian, Michel Bauer, Marie-Claude David, Michel Le Bellac, Kirone Mallick, Vincent Pasquier, Catherine Pépin. I got also much benefit, inspiration and stimulation from shorter discussions with, or from attending talks by, Alain Aspect, Thibault Damour, Nicolas Gisin, Lucien Hardy, Franck Laloë, Theo M. Nieuwenhuizen, Robert Spekkens (although some of them may not be aware of it). Most of theses notes have been elaborated and written at the Institut de Physique théorique of Saclay. I also benefited from the stimulating environment of the Perimeter Institute for Theoretical Physics in Waterloo, ON, and of the inspiring one of the Erwin Schrödinger Institute in Wien, where these lecture notes were achieved. Finally I am very grateful to Christian Caron for giving me the opportunity to refine and polish these notes and to have them published.

Chapter 2
The Standard Formulations of Classical and Quantum Mechanics

I first start by reminders of classical mechanics, probabilities and quantum mechanics, in their usual formulations in theoretical physics. This is mostly very standard material. The last section on reversibility and probabilities in quantum mechanics is a slightly more original presentation of these questions.

2.1 Classical Mechanics

Classical mechanics can be formulated using the Lagrangian formulation or the Hamiltonian formulation. They apply both to non relativistic systems of particles, to fields (like the electromagnetic field) and to relativistic systems. They are valid for closed non-dissipative systems. Macroscopic systems with dissipation, and more generally out of equilibrium open systems must be studied by the tools of statistical mechanics. This is still very active field of research, but out of the scope of this short presentation.

The standard books on classical mechanics are the books by Landau and Lifshitz [72] and the book by A. Arnold [6].

2.1.1 The Lagrangian Formulation

In the Lagrangian formulation, a classical (at this point non-relativistic) system is described by its configuration space $\mathcal{C}$ (a Lagrangian manifold). A point $\mathbf{q} = \{q^i\}_{i=1,N}$ of $\mathcal{C}$ describes an instantaneous configuration of the system. The N coordinates q^i label the N degrees of freedom (d.o.f.) of the system. The state of the system at time t is given by its configuration $\mathbf{q}$ and its velocity $\dot{\mathbf{q}} = \frac{d\mathbf{q}}{dt}$ in configuration space. The dynamics is given by the equations of motion, that take the

© Springer International Publishing Switzerland 2015
F. David, *The Formalisms of Quantum Mechanics*, Lecture Notes in Physics 893,
DOI 10.1007/978-3-319-10539-0_2

form of the Euler-Lagrange equations. They derive from the Lagrange function (the Lagrangian) $L(\mathbf{q}, \dot{\mathbf{q}})$ and read

$$\frac{d}{dt} \frac{\partial L(q, \dot{q})}{\partial \dot{q}(t)} = \frac{\partial L(q, \dot{q})}{\partial q(t)} \tag{2.1}$$

Let us consider the simple textbook case of a non relativistic particle of mass m in a one dimensional space (a line). Its coordinate (position) is denoted q. It is submitted to a conservative force which derives from a potential $V(q)$. The potential is independent of time. The velocity is $\dot{q}(t) = \frac{dq}{dt}$. The dynamics of the particle is given by Newton's equation

$$m \ddot{q}(t) = -\frac{\partial}{\partial q} V(q) \tag{2.2}$$

which is obtained from the Lagrangian

$$L(q, \dot{q}) = \frac{m}{2} \dot{q}^2 - V(q) \tag{2.3}$$

The equation of motion derives from the least action principle. The action $S[q]$ of a general trajectory $q = \{q(t)\}$ starting from q_1 at time t_1 and ending at q_2 at time t_2 is defined as the integral of the Lagrangian

$$S[q] = \int_{t_i}^{t_f} dt \, L(q(t), \dot{q}(t)) \tag{2.4}$$

The classical trajectory q_c, that satisfy the equations of motion, is the trajectory that extremizes the action S under small variations that leave the initial and final points unmoved.

$$q(t) = q_c(t) + \delta q(t), \quad \delta q(t_1) = \delta q(t_2) = 0 \tag{2.5}$$

The stationarity condition

$$q(t)=q_c(t)+\delta q(t), \quad \delta q(t_i)=\delta q(t_f)=0 \quad \Longrightarrow \quad S[q_c + \delta q] = S[q_c] + \mathcal{O}(\delta q^2) \tag{2.6}$$

can be rewritten as the vanishing of the functional derivative of the action

$$\frac{\delta S[q]}{\delta q(t)} = -\frac{d}{dt} \frac{\partial L(q, \dot{q})}{\partial \dot{q}(t)} + \frac{\partial L(q, \dot{q})}{\partial q(t)} = 0 \tag{2.7}$$

which leads to the Euler-Lagrange equations and to Newton's equation (2.2).

   The Lagrangian formalism is valid for most conservative physical systems. For N
particles in d dimensions the configuration space is $N \times d$ dimensional, and a configuration is a $d \times N$-dimensional vector $\mathbf{q} = \{q_a^i; i = 1, d, a = 1, N\}$. The formalism is more generally valid for charged particles in an external magnetic field, for
mechanical systems with constraints and submitted to external forces, for the classical electromagnetic field and for general classical fields, provided that the dynamics
is reversible and non-dissipative (but the dynamics may be non-invariant under time
reversal, think about the well known example of the motion of a charged particle
in a magnetic field). It applies also to relativistic systems and relativistic fields, and
to the space-time of general relativity as well (with a proper treatment of gauge
symmetries, of space and time and of Lorentz and diffeomorphism invariance)
   The Lagrangian formulation is especially well suited to discuss the role and
consequences of symmetries. For instance the relativistic scalar field (the classical
Klein-Gordon field) is described by a classical real field $x \to \phi(x)$ in Minkowski
space-time $x = (x^\mu) = (t, \vec{x}) \in \mathbb{M}^{1,3}$, with metric $h = \text{diag}(-1, 1, 1, 1)$, i.e. space-
time line element $ds^2 = -dt^2 + d\vec{x}^2$. Its Lagrangian density L and its action S are

$$L(\phi, \partial\phi) = -\frac{1}{2}\partial_\mu\phi\partial^\mu\phi - \frac{m^2}{2}\phi^2 \quad , \qquad S[\phi] = \int d^4x \, L(\phi, \partial\phi) \qquad (2.8)$$

and are explicitly invariant under Lorentz and Poincaré transformations (Fig. 2.1).
   It is in this Lagrangian formalism that it is easy to prove the famous Noether's
theorem. This theorem states that to any symmetry (continuous group of invariance)
of the dynamics (the action) is associated a conserved charge (a conserved local
current for a field).
   The "teleological aspect" of the least action principle (the trajectory is defined as
a function of its initial and final conditions) seemed for a long time mysterious and
had to wait for quantum mechanics and its path integral formulation by R. Feynman
to be fully understood. The Lagrangian formalism is thus well suited to discuss path
integral and functional integral quantization.

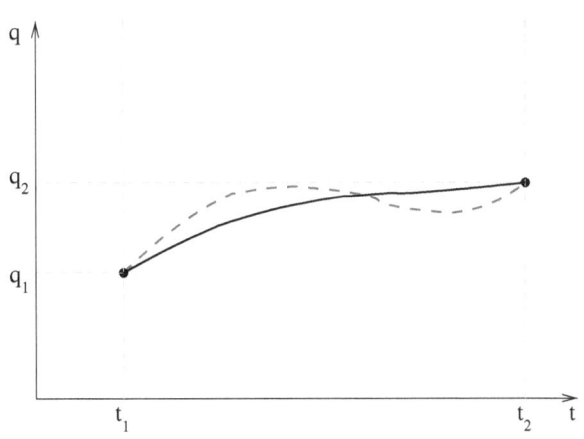

Fig. 2.1 The least action
principle in configuration
space: the classical trajectoire
(*full line*) extremizes the
action $S[q]$. Under a variation
(*dashed line*) $\delta S = 0$. The
initial and final times (t_1, t_2)
and positions (q_1, q_2) are kept
fixed

2.1.2 The Hamiltonian Formulation

2.1.2.1 The Phase Space and the Hamiltonian

The Hamiltonian formulation is in general equivalent to the Lagrangian formulation, but in fact slightly more general. It is well suited to discuss the relation between classical physics and the "canonical quantization" schemes. It allows to discuss in general the structure of the states and of the observables of a classical system, without reference to a specific choice of configuration variables and of configuration space.

For a classical system with n degrees of freedom, the set of possible states of the system is its phase space $\boldsymbol{\Omega}$. A state of the system is now a point $\boldsymbol{x}$ in the phase space $\boldsymbol{\Omega}$ of the system. $\boldsymbol{\Omega}$ is a manifold with even dimension $2n$. The evolution equations are flow equations (first order differential equations in time) in the phase space, and derive from the Hamiltonian function $H(\boldsymbol{x})$.

For the particle in dimension $d = 1$ in a potential there is one degree of freedom, $n = 1$, so that $\boldsymbol{\Omega} = \mathbb{R}^2$ and $\dim(\boldsymbol{\Omega}) = 2$. The two coordinates in phase space are the position q et the momentum p of the particle.

$$\boldsymbol{x} = (q, p) \tag{2.9}$$

The Hamiltonian is

$$H(q, p) = \frac{p^2}{2m} + V(q) \tag{2.10}$$

The equations of motion are the Hamilton equations

$$\dot{p} = -\frac{\partial H}{\partial q} \quad , \quad \dot{q} = \frac{\partial H}{\partial p} \tag{2.11}$$

The relation between the momentum and the velocity $p = m\dot{q}$ is now a dynamical relation. The dissymmetry between q and p comes in fact from the antisymmetric Poisson bracket structure present in the phase space, this will be discussed a bit later.

The Hamilton equations derive also from a variational principle. To find the classical trajectory in phase space such that $q(t_1) = q_1$, $q(t_2) = q_2$ one has to extremize the action functional (the "Hamiltonian action") $\mathcal{S}_H$

$$\mathcal{S}_H[q, p] = \int_{t_1}^{t_2} dt \left(p(t)\dot{q}(t) - H(q(t), p(t)) \right) \tag{2.12}$$

with respect to variations of $q(t)$ and of $p(t)$, $q(t)$ being fixed at the initial and final times $t = t_1$ et t_2, but $p(t)$ being left free at $t = t_1$ and t_2. Indeed, the functional derivatives of $\mathcal{S}_H$ are

$$\frac{\delta \mathcal{S}_H}{\delta q(t)} = -\dot{p}(t) - \frac{\partial H}{\partial q}(q(t), p(t)) \,, \qquad \frac{\delta \mathcal{S}_H}{\delta p(t)} = \dot{q}(t) - \frac{\partial H}{\partial p}(q(t), p(t)) \qquad (2.13)$$

The change of variables $(q, \dot{q}) \rightarrow (q, p)$, of the Lagrangian to the Hamiltonian $L(q, \dot{q}) \rightarrow H(q, p)$ and of the action functionals $S[q, \dot{q}] \rightarrow \mathcal{S}_H[q, p]$ between the Lagrangian and the Hamiltonian formalism correspond to a Legendre transformation with respect to the velocity $\dot{q}$. The velocity $\dot{q}$ and the momentum p are conjugate variables. Indeed one has the relation

$$\mathbf{p} = \frac{\partial L(\mathbf{q}, \dot{\mathbf{q}})}{\partial \dot{\mathbf{q}}} \quad , \quad H(\mathbf{q}, \mathbf{p}) = \mathbf{p}\dot{\mathbf{q}} - L(\mathbf{q}, \dot{\mathbf{q}}) \qquad (2.14)$$

2.1.2.2 Hamilton-Jacobi Equation

For a classical trajectory $q_{cl}(t)$ solution of the equations of motion, the "Hamiltonian" action $\mathcal{S}_H[q_c, p_c]$ and the Lagrangian action $S[q_c]$ are equal. This is not really surprising, this is a property of the Legendre transformation.

Now let us fix the initial time t_1 and the initial position q_1 of the particle. This classical action can be considered now as being a function of the final time $t_2 = t$ and of the final position $q(t_2) = q_2 = q$ of the particle. This function is called the Hamilton-Jacobi action, or the Hamilton function (not to be confused with the Hamiltonian), and let us denote it $\mathcal{S}(q, t) = \mathcal{S}_{HJ}(q, t)$ to be explicit (the initial conditions $q(t_1) = q_1$ being implicit) (Fig. 2.2)

$$\mathcal{S}(q, t) = \mathcal{S}_{HJ}(q, t) = S[q_{cl}] \quad \text{with } q_{cl} \text{ classical solution such that} \quad q(t_2)=q, t_2=t$$

$$\text{and where } t_1 \text{ and } q(t_1) = q_1 \text{ are kept fixed}$$

$$(2.15)$$

Using the equations of motion it is easy to see that the evolution with the final time t of this function $\mathcal{S}(q, t)$ is given by the differential equation

$$\frac{\partial \mathcal{S}}{\partial t} = -H\left(q, \frac{\partial \mathcal{S}}{\partial q}\right) \qquad (2.16)$$

with H the Hamiltonian function. This is a first order differential equation with respect to the final time t. It is called the Hamilton-Jacobi equation.

From this equation on can show that (the initial conditions (t_1, q_1) being fixed) the impulsion p and the total energy E of the particle, expressed as a function of its final position q and of the final time t, are

$$E(q, t) = -\frac{\partial \mathcal{S}}{\partial t}(q, t) \,, \quad p(q, t) = \frac{\partial \mathcal{S}}{\partial q}(q, t) \qquad (2.17)$$

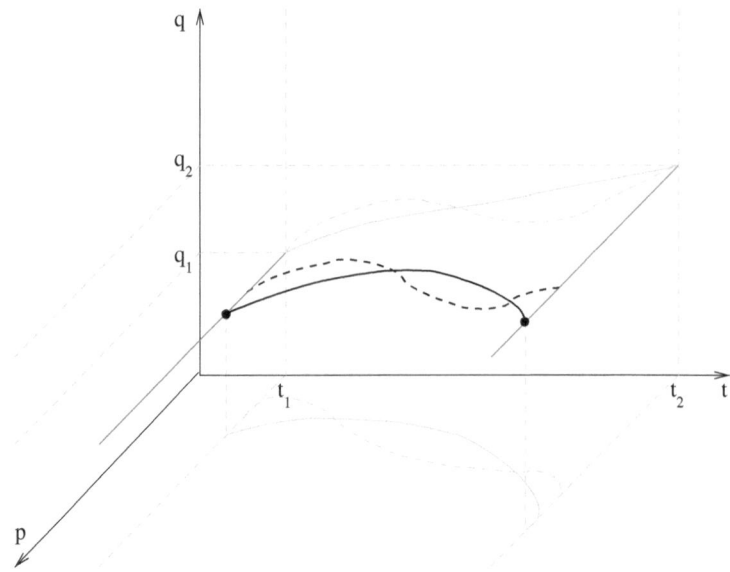

Fig. 2.2 The least action principle in phase space: The classical trajectory in phase space (*full line*) extremizes the action $S_H[q, p]$. The initial and final positions q_1 and q_2 and times t_1 and t_2 are fixed. The initial and final momenta p_1 and p_2 are free. Their actual value is given by the variational principle, and is a function of the initial and final positions and of the initial and final times

These equations extend to the case of systems with n degrees of freedom and of more general Hamiltonians, positions and momenta being now n components vectors

$$\mathbf{q} = \{q^i\} \quad , \qquad \mathbf{p} = \{p_i\} \qquad i = 1, \cdots, n \qquad (2.18)$$

The Hamilton-Jacobi equations are quite important, in particular in the context of the semi-classical limit of quantum mechanics, where the Hamilton-Jacobi action turns to be the "quantum phase" of the quantum particle.

2.1.2.3 Symplectic Manifolds

For the most general case (for example of classical systems with constraints), the Hamiltonian formulation requires the formalism of symplectic geometry. The phase space $\mathbf{\Omega}$ of a system with n degrees of freedom is a manifold with an even dimension $N = 2n$. It is not necessarily the Euclidean space $\mathbb{R}^N$. As we shall see it is only locally like $\mathbb{R}^N$, but it may have a non trivial topology leading to interesting physical effects.

Locally the phase space $\mathbf{\Omega}$ is described by local coordinates $\mathbf{x} = \{x^i, i = 1, 2n\}$. Warning! Now the $2n$ coordinates x_i are coordinate in phase space, not some physical spatial position coordinates in configuration space. Again in general such coordinates systems are not global and must be patched together by some coordinate changes (with properties to be defined later) to get a global description of the phase space.

The Hamiltonian dynamics requires a symplectic structure on $\mathbf{\Omega}$. This symplectic structure allows to define (or amounts to define) the Poisson brackets for the system. $\mathbf{\Omega}$ is said to be a symplectic manifold if it is embodied with an antisymmetric 2-form ω (a degree 2 differential form) which is non-degenerate and closed ($d\omega = 0$). This means that to each point $\mathbf{x} \in \mathbf{\Omega}$ is associated (in the coordinate system $\{x^i\}$ the $2n \times 2n$ antisymmetric matrix (ω_{ij}), that defines the 2-form

$$\omega(\mathbf{x}) = \frac{1}{2}\omega_{ij}(\mathbf{x})\, dx^i \wedge dx^j \qquad (2.19)$$

We use here the standard notations of differential geometry. $dx^i \wedge dx^j$ is the antisymmetric product (exterior product) of the two 1-forms dx^i and dx^j.

The matrix $\omega(\mathbf{x})$ is antisymmetric and non-degenerate,

$$\omega_{ij}(\mathbf{x}) = -\omega_{ji}(\mathbf{x}) \quad , \qquad \det(\omega) \neq 0 \qquad (2.20)$$

This implies that it is invertible. The form ω is moreover closed. This means that its exterior derivative $d\omega$ is zero

$$d\omega(\mathbf{x}) = 0 \qquad (2.21)$$

with

$$d\omega(\mathbf{x}) = \frac{1}{3!}\sum_{i,j,k} \partial_i \omega_{jk}(\mathbf{x})\, dx^i \wedge dx^j \wedge dx^k \qquad (2.22)$$

In term of components this means

$$\forall\, i_1 < i_2 < i_3 \quad , \qquad \partial_{i_1}\omega_{i_2 i_3} + \partial_{i_2}\omega_{i_3 i_1} + \partial_{i_3}\omega_{i_1 i_2} = 0$$

The fact that ω is a differential form means that under a local change of coordinates $\mathbf{x} \to \mathbf{x}'$ (in phase space) the components of the form change as

$$\mathbf{x} \to \mathbf{x}' \quad , \qquad \omega = \omega(\mathbf{x})_{ij}\, dx^i \wedge dx^j = \omega'(\mathbf{x}')_{ij}\, dx'^i \wedge dx'^j$$

that is

$$\omega'(\mathbf{x}')_{ij} = \omega(\mathbf{x})_{kl} \frac{\partial x^k}{\partial x'^i}\frac{\partial x^l}{\partial x'^j}$$

The symplectic form allows an intrinsic definition of the Poisson brackets (see below).

2.1.2.4 Simple Example

For the simple case of a particle on a line, $n = 1$, $\Omega = \mathbb{R}^2$, $\mathbf{x} = (q, p)$, The symplectic form is simply $\omega = dq \wedge dp$. Its components are

$$\omega = (\omega_{ij}) = \begin{pmatrix} 0 & 1 \\ -1 & 0 \end{pmatrix} \qquad (2.23)$$

For a particle in n dimensional space, the configuration space is $\mathbb{R}^n$, the phase space is $\Omega = \mathbb{R}^{2n}$, with coordinates $\mathbf{x} = (q^i, p^i)$, and the symplectic form is $\omega = \frac{1}{2} \sum_i dq^i \wedge dp^i$. It can be written as the block matrix

$$(\omega_{ij}) = \begin{pmatrix} 0 & 1 & 0 & 0 & \cdots \\ -1 & 0 & 0 & 0 & \cdots \\ & & & & \\ 0 & 0 & 0 & 1 & \cdots \\ 0 & 0 & -1 & 0 & \cdots \\ & & & & \\ \vdots & \vdots & \vdots & \vdots & \ddots \end{pmatrix} \qquad (2.24)$$

Darboux's theorem states that for any symplectic manifold (Ω, ω), it is always possible to find locally a coordinate systems such that the symplectic form takes the form (2.24). In such a Darboux coordinate system ω is constant and is a direct sum of antisymmetric symbols. The (q^i, p^j) are said to be local pairs of conjugate variables (the generalization of the conjugate position and momentum variables).

The fact that locally the symplectic form may be written under this generic constant form means that symplectic geometry is characterized by global invariants, not by local ones. This is different from Riemannian geometry, where the metric tensor g_{ij} cannot in general be written as a flat metric $h_{ij} = \delta_{ij}$, and where there are local invariants, such as the scalar curvature R, and many others.

2.1.2.5 Observables, Poisson Brackets

The observables of the system defined by a symplectic phase space (Ω, ω) are identified with the "smooth" real functions from $\Omega \rightarrow \mathbb{R}$ (smooth means "sufficiently regular", at least differentiable and in general C^∞) real functions from $\Omega \rightarrow \mathbb{R}$. The (observed) value of an observable f for the system in the state $\mathbf{x}$ is simply the value of the function $f(\mathbf{x})$. Of course observables may depend also explicitly on the time t.

$$\text{system in state } \mathbf{x} \quad \rightarrow \quad \text{value of } f = f(\mathbf{x}) \qquad (2.25)$$

For two differentiable functions (observables) f and g, their Poisson bracket $\{f, g\}_\omega$ is the function defined by

$$\{f, g\}_\omega(x) = \omega^{ij}(x)\, \partial_i f(x)\, \partial_j g(x) \quad \text{with} \quad \partial_i = \frac{\partial}{\partial x^i} \quad \text{and} \quad \omega^{ij}(x) = \left(\omega^{-1}(x)\right)_{ij} \tag{2.26}$$

the matrix elements of the inverse of the antisymmetric matrix $\omega(x)$ (remember that ω is non degenerate, hence invertible). When no ambiguity is present, the subscript ω will be omitted in the Poisson bracket $\{f, g\}$.

In a canonical local coordinate system (Darboux coordinates) the Poisson bracket takes the standard form

$$\{f, g\} = \sum_i \frac{\partial f}{\partial q^i} \frac{\partial g}{\partial p^i} - \frac{\partial f}{\partial p^i} \frac{\partial g}{\partial q^i} \quad \text{and} \quad \{q^i, p^j\} = \delta_{ij} \tag{2.27}$$

The Poisson bracket is bi-linear, antisymmetric

$$\{f, g\} = -\{g, f\} \tag{2.28}$$

Since it involves only first order derivatives it satisfies the Leibnitz rule (the Poisson bracket acts as a derivation w.r.t to one of its term onto the other one)

$$\{f, gh\} = \{f, g\}h + g\{f, h\} \tag{2.29}$$

The fact that the symplectic form is closed $d\omega = 0$ is equivalent to the Jacobi identity

$$\{f, \{g, h\}\} + \{g, \{h, f\}\} + \{h, \{f, g\}\} = 0 \tag{2.30}$$

Finally, in general coordinates, the knowledge of the Poisson bracket $\{\ ,\ \}$ is equivalent to the knowledge of the symplectic form ω via its inverse ω^{-1}

$$\{x^i, x^j\} = \omega^{ij}(\mathbf{x}) \tag{2.31}$$

2.1.2.6   The Dynamics and Hamiltonian Flows

The purpose of this formalism is of course to describe the dynamics in phase espace. It is generated by an Hamiltonian function H. The Hamiltonian is a real "smooth" (at least differentiable) function on phase space $\Omega \rightarrow \mathbb{R}$, like the observables. The state of the system $\mathbf{x}(t)$ evolves with time according to the Hamilton equation

$$\dot{\mathbf{x}}(t) = \{\mathbf{x}(t), H\} \tag{2.32}$$

In terms of coordinates this reads

$$\dot{x}^i(t) = \omega^{ij}(\mathbf{x}(t))\, \partial_j H(\mathbf{x}(t)) \tag{2.33}$$

The Hamilton equations involve the Poisson Bracket and is covariant under local changes of coordinates in phase space. They are flow equations of the form

$$\dot{x}^i(t) = F^i(\mathbf{x}(t)) \tag{2.34}$$

but the vector field $F^i = \omega^{ij}\partial_j H$ is very special and derives from the Hamiltonian H. The time flow, i.e. the application $\phi\colon \Omega \times \mathbb{R} \to \Omega$ is called the Hamiltonian flow associated to H. The evolution functions $\phi_t(\mathbf{x})$ defined by

$$\mathbf{x}(t=0) = \mathbf{x} \quad \Longrightarrow \quad \mathbf{x}(t) = \phi_t(\mathbf{x}) \tag{2.35}$$

form a group of transformations (as long as H is independent of the time)

$$\phi_{t_1+t_2} = \phi_{t_1} \circ \phi_{t_2} \tag{2.36}$$

More generally, let us consider a (time independent) observable f (a function on Ω). The evolution of the value of the observable f for a dynamical state $\mathbf{x}(t)$, $f(\mathbf{x},t) = f(\mathbf{x}(t))$ where $\mathbf{x}(t) = \phi_t(\mathbf{x})$, obeys the equation

$$\frac{\partial f(\mathbf{x},t)}{\partial t} = \{f, H\}(\mathbf{x}(t)) \tag{2.37}$$

where the r.h.s. is the Poisson bracket of the observable f and the Hamiltonian H. In particular (when H is independent of t) the energy $E(t) = H(\mathbf{x}(t))$ is conserved

$$\frac{\partial E(\mathbf{x},t)}{\partial t} = 0 \tag{2.38}$$

2.1.2.7 The Liouville Measure

The symplectic form ω defines an invariant volume element $d\mu$ on phase space Ω.

$$d\mu(\mathbf{x}) = \omega^n = \prod_{i=1}^{2n} dx^i \, |\omega|^{1/2} \quad , \quad |\omega| = |\det(\omega_{ij})| \tag{2.39}$$

This is the Liouville measure on Ω. This measure is invariant under all the Hamiltonian flows, and is in fact the only local invariant.

2.1.2.8 A Less Trivial Example: The Classical Spin

The simplest example of a system with a non trivial phase space is the classical spin (the classical top with constant total angular momentum). The states of the spin are labelled by unit 3-components vector $\vec{n} = (n_1, n_2, n_3)$, $|\vec{n}| = 1$ (the direction of the angular momentum). Thus the phase space is the two-dimensional unit sphere and is now compact (hence different from $\mathbb{R}^2$)

$$\Omega = \mathcal{S}_2$$

The classical precession equation

$$\frac{d\vec{n}}{dt} = \vec{B} \times \vec{n}$$

can be written in Hamiltonian form. $\vec{B}$ is a vector in $\mathbb{R}^3$, possibly a 3-component vector field on the sphere depending on $\vec{n}$.

There is a symplectic structure on Ω. It is related to the natural complex structure on $\mathcal{S}_2$ (the Riemann sphere). The Poisson bracket of two functions f and g on $\mathcal{S}_2$ is defined as

$$\{f, g\} = (\vec{\nabla} f \times \vec{\nabla} g) \cdot \vec{n} .$$

The gradient field $\vec{\nabla} f$ of a function f on the sphere is a vector field tangent to the sphere, so $\vec{\nabla} f \times \vec{\nabla} g$ is normal to the sphere, hence collinear with $\vec{n}$. In spherical coordinates

$$\vec{n} = (\sin\theta \cos\phi, \sin\theta \sin\phi, \cos\theta)$$

the Poisson bracket is simply

$$\{f, g\} = \frac{1}{\sin\theta} \left(\frac{\partial f}{\partial \theta} \frac{\partial g}{\partial \phi} - \frac{\partial g}{\partial \theta} \frac{\partial f}{\partial \phi} \right)$$

Admissible local Darboux coordinates $\mathbf{x} = (x^1, x^2)$ such that $\omega = dx^1 \wedge dx^2$ must be locally orthogonal, area preserving mappings $\mathbb{R}^2 \to \mathcal{S}_2$. Examples are the "action-angle" variables (the Lambert cylindrical equal-area projection)

$$\mathbf{x} = (\cos\theta, \phi)$$

and the plane coordinates (the Lambert azimuthal equal-area projection).

$$\mathbf{x} = (2\sin(\theta/2)\cos\phi, 2\sin(\theta/2)\sin\phi)$$

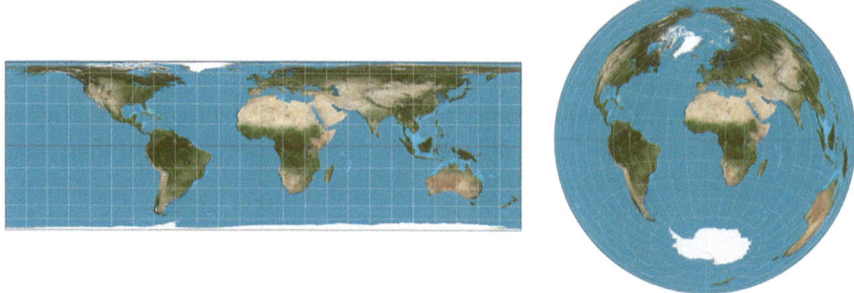

Fig. 2.3 The Lambert cylindrical and azimuthal coordinates on the sphere (from Wikimedia Commons)

The Hamiltonian which generates the precession dynamics is simply (for constant $\vec{B}$) (Fig. 2.3)

$$H = \vec{B} \cdot \vec{n}$$

2.1.2.9 Statistical States, Distribution Functions, the Liouville Equation

We now consider statistical ensembles. If one has only partial information on the state of the system (for instance if one consider an instance of the system taken at random in a big ensemble, or if we know only the value of some of its observables) this information is described by a statistical state (also called mixed state or statistical ensemble) φ. Mixed states are described by probability distributions on the phase space Ω,

$$d\rho_\varphi(\mathbf{x}) = d\mu(\mathbf{x})\, \rho_\varphi(\mathbf{x}) \tag{2.40}$$

with $d\mu(\mathbf{x})$ the Liouville measure and $\rho_\varphi(\mathbf{x})$ the probability density, a non negative distribution (function) such that

$$\rho_\varphi(\mathbf{x}) \geq 0 \quad, \qquad \int_\Omega d\mu(\mathbf{x})\, \rho_\varphi(\mathbf{x}) = 1 \tag{2.41}$$

For a system in a statistical state φ the expectation for an observable f (its expectation value, i.e. its mean value if we perform a large number of measurements of f on independent systems in the same ensemble φ) is

$$\langle f \rangle_\varphi = \int_\Omega d\mu(\mathbf{x})\, \rho_\varphi(\mathbf{x})\, f(\mathbf{x}) \tag{2.42}$$

When the system evolves according to the Hamiltonian flow ϕ_t generated by an Hamiltonian H, the statistical state evolves with time $\varphi \to \varphi(t)$, as well as the distribution function $\rho_\varphi \to \rho_{\varphi(t)}$. φ being the initial state of the system at time $t = 0$, we denote this distribution function

$$\rho_{\varphi(t)}(\mathbf{x}) = \rho_\varphi(\mathbf{x}, t) \tag{2.43}$$

It is related to the initial distribution function at time $t = 0$ by

$$\rho_\varphi(\mathbf{x}(t), t) = \rho_\varphi(\mathbf{x}) \quad , \qquad \mathbf{x}(t) = \phi_t(\mathbf{x}) \tag{2.44}$$

(the Liouville measure is conserved by the Hamiltonian flow). Using the evolution equation for $\mathbf{x}(t)$ (2.33), one obtains the Liouville equation as the evolution equation for the distribution function $\rho_\varphi(\mathbf{x}, t)$

$$\frac{\partial}{\partial t} \rho_\varphi = \{H, \rho_\varphi\} \tag{2.45}$$

The entropy of the statistical state φ is given by the Boltzmann-Gibbs formula (with $k_B = 1$)

$$S = -\int_\Omega d\mu(\mathbf{x}) \, \rho_\varphi(\mathbf{x}) \, \log(\rho_\varphi(\mathbf{x})) \tag{2.46}$$

Of course when the state of the system is fully determined, it is a "pure state" $\mathbf{x}_0$ ($\varphi_{\text{pure}} = \mathbf{x}_0$) the distribution function is a Dirac measure $\rho_{\text{pure}}(\mathbf{x}) = \delta(\mathbf{x} - \mathbf{x}_0)$.

2.1.2.10 Canonical Transformations

Hamiltonian flows are canonical transformations, i.e. examples of (bijective) mappings $\mathcal{C} \ \Omega \ \to \ \Omega$ that preserve the symplectic structure. Denoting $\mathbf{X} = \mathcal{C}(\mathbf{x})$ the image of the point $\mathbf{x} \in \Omega$ by the canonical transformation $\mathcal{C}$, this means simply that the symplectic form ω^* defined by

$$\omega^*(x) = \omega(X) \tag{2.47}$$

is equal to the original form $\omega^* = \omega$. ω^* is called the pullback of the symplectic form ω by the mapping $\mathcal{C}$.

Canonical transformations preserve the Poisson brackets. Let f and g be two observables (functions $\Omega \to \mathbb{R}$ and $F = f \circ \mathcal{C}^{-1}$ and $G = g \circ \mathcal{C}^{-1}$ their transform by the transformation $\mathcal{C}$

$$f(\mathbf{x}) = F(\mathbf{X}) \quad , \qquad g(\mathbf{x}) = G(\mathbf{X}) \tag{2.48}$$

$\mathcal{C}$ is a canonical transformation if

$$\{f, g\}_\omega = \{F, G\}_\omega \tag{2.49}$$

Taking for f and g the coordinate change $x^i \to X^i$ itself, canonical transformations are change of coordinates such that

$$\{X^i, X^j\} = \{x^i, x^j\} \tag{2.50}$$

Canonical transformations are very useful tools in classical mechanics. They are the classical analog of unitary transformations in quantum mechanics.

In the simple example of the classical spin, the canonical transformations are the smooth area preserving diffeomorphisms of the two dimensional sphere.

2.1.2.11 Along the Hamiltonian Flows

As an application, one can treat the Hamiltonian flow ϕ_t as a time dependent canonical transformation (a change of reference frame) and consider the dynamics of the system in this new frame, that evolves with the system. In this new coordinates, that we denote $\bar{\mathbf{x}} = \{\bar{x}^i\}$, if at time $t = 0$ the system is in the initial state $\bar{\mathbf{x}} = \mathbf{x}_0$, at time t it is still in the same state $\bar{\mathbf{x}}(t) = \mathbf{x}_0$. while the observables f become time dependent.

Indeed, starting from a time independent observable f, $\mathbf{x} \to f(\mathbf{x})$, let us denote $\bar{f}$ the time dependent observable in the new frame,

$$\bar{f}(\bar{\mathbf{x}}, t) = f(\mathbf{x}(t)) \qquad \text{with} \quad \mathbf{x}(t) = \phi_t(\bar{\mathbf{x}}) \tag{2.51}$$

It describes how the observable f evolves with t, as a function of the initial state $\bar{\mathbf{x}}$ at $t = 0$. Its evolution is given by

$$\frac{\partial \bar{f}}{\partial t} = \{\bar{f}, H\} \tag{2.52}$$

This change of frame corresponds to changing from a representation of the dynamics by an evolution of the states, the observables being time independent, to a representation where the states do not evolve, but where there observables depend on time. This is the analog for Hamiltonian dynamics to what is done in fluid dynamics: going from the Eulerian specification (the fluid moves in a fixed coordinate system) to the Lagrangian specification (the coordinate system moves along the fluid). These two representations are of course the classical analog of the Schrödinger picture (vector states evolves, operators are fixed) and of the Heisenberg picture (vector states are fixed, operators depend on time) in quantum mechanics.

2.1.3 The Algebra of Classical Observables

One can adopt a more abstract point of view. It will be useful for quantum mechanics.

2.1.3.1 Functions as Elements of a Commutative C^*-Algebra

The real functions (continuous, with compact support) on phase space $f; \Omega \to \mathbb{R}$ form a commutative algebra $\mathcal{A}$ with the standard addition and multiplication laws.

$$(f + g)(x) = f(x) + g(x) \quad , \qquad (fg)(x) = f(x)g(x) \tag{2.53}$$

Statistical states (probability distributions on Ω) can be viewed as normalized positive linear forms φ on $\mathcal{A}$, i.e. applications $\mathcal{A} \to \mathbb{R}$ such that

$$\varphi(\alpha f + \beta g) = \alpha \varphi(f) + \beta \varphi(g) \text{ , with } \alpha, \beta \in \mathbb{R} \tag{2.54}$$

$$\varphi(|f|^2) \geq 0 \,, \quad \varphi(1) = 1 \tag{2.55}$$

The sup norm or $\mathcal{L}^\infty$ norm, is defined on $\mathcal{A}$ as

$$\|f\|^2 = \sup_{x \in \Omega} |f(x)|^2 = \sup_{\varphi \text{ states}} \varphi(|f(x)|^2) \tag{2.56}$$

It has clearly the following properties (extending $\mathcal{A}$ from the algebra of real function to the algebra of complex functions)

$$\|f\| = \|f^*\| \,, \quad \|fg\| \leq \|f\| \, \|g\| \,, \quad \|ff^*\| = \|f\|^2 \tag{2.57}$$

and $\mathcal{A}$ is complete under this norm (the limit of a convergent series in $\mathcal{A}$ belongs to $\mathcal{A}$). This makes the algebra $\mathcal{A}$ a mathematical object denoted a commutative C*-algebra.

A famous theorem by Gelfand and Naimark states that the reciprocal is true. Any commutative C*-algebra $\mathcal{A}$ is isomorphic to the algebra $C(X)$ of the continuous functions on some topological (locally compact) space X. This seems a somehow formal result (the space X and its topology may be quite wild, and very far from a regular smooth manifold). What is important is that a mathematical object (here a topological space X) can be defined intrinsically (by its elements x) or equivalently by the abstract properties of some set of functions over this object (here the commutative algebra of observables). This modern point of view in mathematics (basically this idea is at the basis of the category formalism) is also important in modern physics, as we shall see later in the quantum case.

Technically, the proof is simple. Starting from some X and $\mathcal{A} = C(X)$, to any element $x \in X$, one associate the subalgebra $\mathcal{I}_x$ of all functions that vanish at x

$$\mathcal{I}_x : \{f \in \mathcal{A}; \ f(x) = 0\} \tag{2.58}$$

The $\mathcal{I}_x$ are the maximal ideals of $\mathcal{A}$, (left-)ideals $\mathcal{I}$ of an algebra $\mathcal{A}$ being subalgebras of $\mathcal{A}$ such that $x \in \mathcal{I}$ and $y \in \mathcal{A}$ implies $xy \in \mathcal{I}$. It is easy to show that the set $\mathcal{X}$ of the maximal ideals of $\mathcal{A} = C(X)$ is isomorphic to X, and that $\mathcal{A}/\mathcal{I}_x = \mathbb{C}$ the target space. Reciprocally, to any commutative C^* algebra $\mathcal{A}$ one can associate the set of its maximal ideas $\mathcal{X}$, show that it is locally compact and that

$$\mathcal{X} = \{\text{maximal ideals of } \mathcal{A}\} \iff \mathcal{A} = C(\mathcal{X}) \tag{2.59}$$

2.1.3.2 Observable as Elements of a Commutative Poisson Algebra

For the Hamiltonian systems, the algebra of observables $\mathcal{A}$ of (now C^∞, i.e. differentiable) functions on the phase space Ω is equipped with an additional product, the antisymmetric Poisson bracket $\{\cdot, \cdot\}$ that satisfy (2.28)–(2.30). This algebra $\mathcal{A}$, with its three laws (addition, multiplication, Poisson bracket) is now a commutative Poisson algebra. Poisson algebras may be non-commutative.

The most general formulation for classical Hamiltonian dynamics is that of Poisson manifolds and of its associated commutative Poisson algebra. Poisson manifolds are manifolds (phase space) embodied with a Poisson bracket, but this is a more general formulation that symplectic manifolds, since it encompasses special situations where the Poisson bracket is degenerate, and does not define a symplectic structure. They are useful for some general situation, in particular when studying the effective classical dynamics that emerge in some semiclassical limit from a fully quantum dynamics. Poisson manifolds can in general be split (foliated) into "symplectic leaves" embodied with a well defined induced symplectic structure.

The fact that in classical mechanics dynamics are given by Hamiltonian flows on a phase space which is a symplectic or a Poisson manifold can be somehow justified at the classical level. One has to assume that the possible dynamics are given by flows equations generated by some smooth vector fields, that these flows are generated by conserved quantities (Hamiltonians) and that the dynamics are covariant under change of frames generated by these flows (existence and invariance of canonical transformations).

However a full understanding and justification of classical Hamiltonian dynamics comes from quantum mechanics. Indeed, the Poisson bracket structure is the "classical limit" of the Lie algebra structure of the commutators of quantum observables (operators) in quantum mechanics, and the canonical transformations are the classical version of the unitary transformations in the Hilbert space of pure states.

2.2 Probabilities

Probabilities are an important aspect of classical physics[1] and are one of the key components of quantum physics, since the later is intrinsically probabilistic. Without going into any details and much formalism, I think it is important to recall the two main ways to consider and use probabilities in mathematics, statistics, physics and natural sciences: the frequentist point of view and the Bayesian point of view. At the level considered here, these are different point of views on the same mathematical formalism, and on its use. As we shall see, in some sense quantum mechanics forces us to treat them on the same footing. There are of course many different, more subtle and more precise mathematical as well as philosophical points of view on probability theory. I shall not enter in any discussion about the merits and the consistency of objective probabilities versus subjective probabilities.

Amongst many standard references on the mathematical formalism of probability, there is the book by Kolmogorov [68], and the book by Feller [43]. See also the quick introduction for and by a physicist by M. Bauer (in French) [12]. References on Bayesian probabilities are the books by de Finetti [34], by Jaynes [65] and the article by Cox [30].

2.2.1 The Frequentist Point of View

The frequentist point of view is the most familiar and the most used in statistical physics, dynamical systems, as well as in mathematics (it is at the basis of the formulation of modern probability theory from the beginning of twentieth century, in particular for the Kolmogorov axiomatic formulation of probabilities). Roughly speaking, probabilities represent a measure of ignorance on the state of a system, coming for instance from: uncertainty on its initial state, uncertainty on its dynamical evolution due to uncertainty on the dynamics, or high sensibility to the initial conditions (chaos). Then probabilities are asymptotic frequencies of events (measurements) if we repeat observations on systems prepared by the same initial procedure. More precisely, one has a set Ω of samples (the sample space), a σ-algebra $\mathcal{F}$ of "measurable" subsets of the sample space Ω, and a measure P on $\mathcal{F}$ (equivalently a probability measure μ on Ω). This probability measure is a priori given. Thus to a subset $E \in \mathcal{F}$ is associated an event E and the probability for this event to happen (its expectation)

$$P(E) = \mathbb{E}[x \in E] = \int_{\Omega} \mu(dx)\, \chi_E(x) \tag{2.60}$$

---

[1]Probability theory appeared and developed in parallel with classical physics, with important contributors in both fields, from Pascal, Bernoulli, and Laplace to Poincaré and Kolmogorov.

2.2.2 The Bayesian Point of View

The so called Bayesian point of view is somehow broader, and of use in statistics, game theory, economy, but also in experimental sciences. It is also closer to the initial formulations of probabilities (or "chance") in the eighteenth and nineteenth centuries. It has been reviewed by statisticians like de Finetti or Jaynes (among others) in the twentieth century.

Probabilities are considered as qualitative estimates for the "plausibility" of some proposition (it can be the result of some observation), given some "state of knowledge" on a system.

$$P(A|C) = \text{plausibility of } A, \text{ knowing } C \qquad (2.61)$$

In particular, one considers the "priors"

$$P(C) = P(\,|C) \qquad (2.62)$$

These probabilities (degree of plausibility) must satisfy rules that are constrained by logical principles, and which turns out to be the rules of probability theory. This is the so called objectivist point of view (objective probabilities), where the degree of plausibility must be established by a "rational agent" from its knowledge of the system. Another (more controversial) point of view is the "subjectivist point of view" (subjective probabilities) where the probabilities $P(A)$ correspond simply to the "degree of personal belief" of the propositions by the agent. In the former usually the priors are constrained by (for instance) some a priori assumed symmetry principle. The difference between the objective and subjective probabilities will be what are the allowed rules for initial choices for the priors. Note also that the concept of Bayesian probabilities is not accepted by everybody, and may be mathematically problematic when dealing with continuous probability measures.

2.2.3 Conditional Probabilities

The basic rules are the same in the different formulations. A most important concept is conditional probabilities $P(A|B)$ (the probability of A, B being given), and the Bayes relation for the conditional probabilities

$$P(A|B) = \frac{P(B|A)P(A)}{P(B)} \qquad (2.63)$$

where $P(A)$ and $P(B)$ are the initial probabilities for A and B (the priors), and $P(A|B)$ and $P(B|A)$ the conditional probabilities.

Fig. 2.4 Venn representation of the conditional probabilities formula

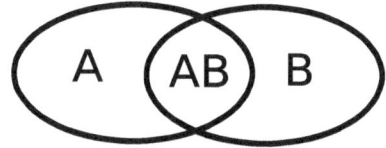

Frequentist In the frequentist formulation $P(A|B)$ is the frequency of A, once we have selected the samples such that B is true. Bayes formula has the simple representation with Venn diagrams in the set of samples (Fig. 2.4)

Bayesian In the Bayesian formulation (see for instance the book by Jaynes), one may consider every probabilities as conditional probabilities. For instance $P_C(A) = P(A|C)$, where the proposition C corresponds to the "prior knowledge" that leads to the probability assignment $p_C(A)$ for A (so P_C is the probability distribution). If AB means the proposition "A and B" ($A \wedge B$ or $A + B$), Bayes formula follows from the "product rule"

$$P(AB|C) = P(A|BC)P(B|C) = P(B|AC)P(A|C) \qquad (2.64)$$

Its meaning is the following: given C, if one already knows the plausibility for AB being true ($P(AB|C)$), and the plausibility for B being true (the prior $P(B|C)$), then (2.64) tells us how one should modify the plausibility for A of being true, if we learn that B is moreover true ($P(A|BC)$). Together with the "sum rule"

$$P(A|C) + P(\neg A|C) = 1 \qquad (2.65)$$

($\neg A$ is the negation of A), these are the basic rules of probability theory in this framework.

2.3 Quantum Mechanics: The "Canonical Formulation"

Let me now recall the so called "canonical formalism" of quantum mechanics. By "standard formalism" I mean nothing but the typical or standard presentation of the formalism, as it is given (with of course many variants) in most textbooks, prior to (or without) the discussions on the significance and the possible interpretations of the formalism. It relies on the formalism of Hilbert spaces, vector states and wave functions, on the "correspondence principle" when discussing the quantization of a classical (usually non-relativistic) system, and aims of course at physical and calculational understanding and efficiency[2] when discussing physical systems and experiments.

---

[2]Looking for efficiency and operability does not mean adopting the (in)famous "shut up and calculate" stance, an advice often but falsely attributed to R. Feynman.

There is of course an enormous number of good books on quantum mechanics and quantum field theory. Among the very first books on quantum mechanics, those of P.A. Dirac [38] and J. von Neumann [105] ([106] for the English traduction of 1955) are still very useful and valuable. Some very good modern books on quantum mechanics with a contemporary view and a treatment of the recent developments are the latest edition of the standard by Cohen-Tanoudji et al. [27], the book by M. le Bellac [73], and the book by Auletta, Fortunato and Parisi [9]. Let me also quote at that stage the book by A. Peres [86], although it is much more focused on the conceptual aspects.

Some standard modern references on quantum field theory are the books by J. Zinn-Justin [113], by S. Weinberg [110] and the book by A. Zee [112] (in a very different relaxed style). Reference more oriented towards mathematical physics will be given later.

Amongst the numerous book and articles on the questions of the foundation and the interpretation of quantum mechanics, very good references are the encyclopedic and balanced review by Auletta [8], and the more recent and shorter book by F. Laloe [71] (see also [69, 70]). More will be given later.

2.3.1 Principles

I give here one of the variants of the canonical formalism, without any justification.

2.3.1.1 Pure States and Hilbert Space

The phase space Ω of classical mechanics is replaced by the complex Hilbert space $\mathcal{H}$ of pure states. Elements of $\mathcal{H}$ (vectors) are denoted ψ or $|\psi\rangle$ ("kets" in Dirac notations). The scalar product of two vectors ψ and ψ' in $\mathcal{H}$ is denoted $\psi^* \cdot \psi'$ or $\langle \psi | \psi' \rangle$. The $\psi^* = \langle \psi |$ are the "bra" and belong to the dual $\mathcal{H}^*$ of $\mathcal{H}$. Note that in the mathematical literature the scalar product is often noted in the opposite order $\langle \psi | \psi' \rangle = \psi' \cdot \psi^*$. We shall stick to the physicists notations.

Pure quantum states are rays of the Hilbert space, i.e. one dimensional subspaces of $\mathcal{H}$. They correspond to unit norm vectors $|\psi\rangle$, such that $\|\psi\|^2 = \langle \psi | \psi \rangle = 1$, and modulo an arbitrary unphysical (unobservable) phase $|\psi\rangle \simeq e^{i\theta}|\psi\rangle$. This normalization condition comes of course from the Born rule (see below).

Of course a consequence of this principle is that any complex linear combination of two states $|\alpha\rangle$ and $|\beta\rangle$, $|\psi\rangle = a|\alpha\rangle + b|\beta\rangle$, corresponds also[3] a state of the system. This is the so-called superposition principle.

---

[3]At least for finite dimensional and simple cases of infinite dimensional Hilbert spaces, see the discussion on superselection sectors.

The complex structure of the Hilbert space is a crucial feature of quantum mechanics. It is reflected in classical mechanics by the almost complex and symplectic structures of the classical phase space.

2.3.1.2 Observables and Operators

The physical observables A are the self-adjoint operators on $\mathcal{H}$ (Hermitian or symmetric operators), such that $A = A^\dagger$, where the conjugation is defined by $\langle A^\dagger \psi' | \psi \rangle = \langle \psi' | A\psi \rangle$. Note that the conjugation $A^\dagger$ is rather denoted A^* in the mathematical literature, and in some sections we shall use this notation, when dealing with general Hilbert spaces not necessarily complex.

The operators on $\mathcal{H}$ form an associative, but non commutative complex operator algebra. Any set of commuting self-adjoint operators $\{A_i\}$ corresponds to a family of classically compatible observables, which in principle can be measured independently.

2.3.1.3 Measurements, Probabilities and the Born Rule

The outcome of the measurement of an observable A on the system in a state ψ is in general not deterministic. Quantum mechanics give only probabilities for the outcomes, and in particular the expectation value of the outcomes $\langle A \rangle_\psi$. This expectation value is given by the Born rule

$$\langle A \rangle_\psi = \langle \psi | A | \psi \rangle = \langle \psi | A\psi \rangle \tag{2.66}$$

For compatible (commuting) observables the probabilities of outcome obey the standard rule of probabilities and these measurements can be repeated and performed independently.

This implies (or is equivalent to state) that the possible outcomes of the measurement of A must belong to the spectrum of A, i.e. can only equal the eigenvalues of A (I consider the simple case where A has a discrete spectrum). Moreover the probability $p_i(\psi)$ to get as outcome the eigenvalue a_i, denoting $|i\rangle$ the corresponding eigenvector, is the modulus squared of the probability amplitude $\langle i | \psi \rangle$

$$p_i(\psi) = \frac{\text{probability of outcome of } A \to a_i}{\text{if the system is in the state } |\psi\rangle} = |\langle i | \psi \rangle|^2$$

A very important consequence (or feature) is that quantum measurements are intrinsically irreversible processes. Let us consider ideal measurements (non destructive measurements), i.e. measurement operations which can be repeated quasi-instantaneously on quantum systems and when repeated, give always the same result. If the system is initially in a state ψ), if one performs an ideal measurement of A with outcome a_i, and if there is single eigenvector $|i\ )$ associated to this eigenvalue a_i of A, this implies that the system must be considered to be in the associated eigenstate $|i\rangle$. If the eigenspace V_i associated to the eigenvalue a_i is a higher dimensional subspace $V_i \subset \mathcal{H}$, then the system must be in the projected state $|\psi_i\rangle = P_i|\psi\rangle$, with P_i the orthogonal projector onto V_i. This is the projection postulate.

For observables with a continuous spectrum and non-normalizable eignestates, the theory of ideal measurements involves more mathematical rigor and the use of spectral theory.

At that stage I do not discuss what it means to "prepare a system in a given state", what "represents" the state vector, what is really a measurement process (the problem of quantum measurement) and what means the projection postulate. We shall come back to some of these questions along the course.

2.3.1.4 Unitary Dynamics

For a closed system, the time evolution of the states is linear and it must preserve the probabilities, hence the scalar product $\langle.|.\rangle$. Therefore is given by unitary transformations $U(t)$ such that $U^{-1} = U^\dagger$. Again if the system is isolated the time evolution form a multiplicative group acting on the Hilbert space and its algebra of observables, hence it is generated by an Hamiltonian self-adjoint operator H

$$U(t) = \exp\left(\frac{t}{i\hbar}H\right)$$

The evolution equations for states and observables are discussed below.

2.3.1.5 Multipartite Systems

Assuming that it is possible to perform independent measurements on two independent (causally) subsystems S_1 and S_2 implies (at least in the finite dimensional case) that the Hilbert space $\mathcal{H}$ of the states of the composite system $S = $ "$S_1 \cup S_2$" is the tensor product of the Hilbert spaces $\mathcal{H}_1$ and $\mathcal{H}_2$ of the two subsystems.

$$\mathcal{H} = \mathcal{H}_1 \otimes \mathcal{H}_2$$

This implies the existence for the system S of generic "entangled states" between the two subsystems

$$|\Psi\rangle = c|\psi\rangle_1 \otimes |\phi\rangle_2 + c'|\psi'\rangle_1 \otimes |\phi'\rangle_2$$

Entanglement is one of the most important feature of quantum mechanics, and has no counterpart in classical mechanics. It is entanglement that leads to many of the counter-intuitive features of quantum mechanics, but it leads also to many of its interesting aspects and to some of its greatest successes.

Finally let me insist on this point: the fact that one can treat parts of a physical system as independent subparts is not obvious. One may ask what is the property that characterizes this fact that two systems are independent, in the sense that they are causally independent. In fact two parts of a system can be considered as independent if all the physical observables relative to the first part commutes with those relative to the second one. This can be properly understood only in the framework of special relativity, and is deeply related to the concept of locality and causality in relativistic quantum theories, namely in quantum field theories.

2.3.1.6 Correspondence Principe, Canonical Quantization

The correspondence principle has been very important in the elaboration of quantum mechanics. Here by correspondence principle I mean that when quantizing a classical system, often one can associate to canonically conjugate variables (q_i, p_i) self-adjoint operators (Q_i, P_i) that satisfy the canonical commutation relations

$$\{q_i, p_i\} = \delta_{ij} \implies [Q_i, P_j] = i\hbar\delta_{ij} \tag{2.67}$$

and to take as Hamiltonian the operator obtained by replacing in the classical Hamiltonian the variables (q_i, p_i) by the corresponding operators.

For instance, for the particle on a line in a potential, one takes as (Q, P) the position and the momentum and for the Hamiltonian

$$H = \frac{P^2}{2m} + V(Q) \tag{2.68}$$

The usual explicit representation is, starting from the classical position space $\mathbb{R}$ as configuration space, to take for Hilbert space the space of square integrable functions $\mathcal{H} = \mathcal{L}^2(\mathbb{R})$, the states $|\psi\rangle$ correspond to the wave functions $\psi(q)$, and the operators are represented as

$$Q = q \quad , \qquad P = \frac{\hbar}{i}\frac{\partial}{\partial q} \tag{2.69}$$

Of course the value of the wave function $\psi(q)$ is simply the scalar product of the state $|\psi\rangle$ with the position eigenstate $|q\rangle$

$$\psi(q) = \langle q|\psi\rangle \quad , \qquad Q|q\rangle = q|q\rangle \tag{2.70}$$

(a proper mathematical formulation involving the formalism of ringed Hilbert spaces).

2.3.2 Representations of Quantum Mechanics

The representation of states and observables as vectors and operators is invariant under global unitary transformations (the analog of canonical transformations in classical mechanics). These unitary transformations may depend on time. Therefore there are different representations of the dynamics in quantum mechanics. I recall the two main representations.

2.3.2.1 The Schrödinger Picture

It is the most simple, and the most used in non relativistic quantum mechanics, in canonical quantization and is useful to formulate the path integral. In the Schrödinger picture the states ψ (the kets $|\psi\rangle$) evolve with time and are noted $\psi(t)$. The observables are represented by time independent operators. The evolution is given by the Schrödinger equation

$$i\hbar \frac{d\psi}{dt} = H\psi \tag{2.71}$$

The expectation value of an observable A measured at time t for a system in the state ψ is thus

$$\langle A \rangle_{\psi(t)} = \langle \psi(t)|A|\psi(t)\rangle \tag{2.72}$$

The evolution operator $U(t)$ is defined by

$$\psi(t=0) = \psi_0 \quad \rightarrow \quad \psi(t) = U(t)\psi_0 \tag{2.73}$$

It is given by

$$U(t) = \exp\left(\frac{t}{i\hbar}H\right) \tag{2.74}$$

and obeys the evolution equation

$$i\hbar \frac{d}{dt}U(t) = H\ U(t) \quad ; \quad U(0) = \mathbf{1} \tag{2.75}$$

This generalizes easily to the case where the Hamiltonian depends explicitly of the time t. Then

$$i\hbar \frac{d}{dt}U(t,t_0) = H(t)\ U(t,t_0) \quad ; \quad U(t_0,t_0) = \mathbf{1} \tag{2.76}$$

and

$$U(t, t_0) = T \left[\exp \left(\frac{1}{i\hbar} \int_{t_0}^{t} dt\, H(t) \right) \right] = \sum_{k=0}^{\infty} (i\hbar)^{-k} \int_{t_0 < t_1 < \cdots < t_k < t} dt_1 \cdots dt_k \; H(t_k) \cdots H(t_1)$$

(2.77)

where T means the time ordered product (more later).

2.3.2.2 The Heisenberg Picture

This representation is the most useful in relativistic quantum field theory. It is in fact the best mathematically fully consistent formulation, since the notion of state in more subtle, in particular it depends on the reference frame. It is required for building the relation between critical systems and Euclidean quantum field theory (statistical field theory).

In the Heisenberg representation, the states are redefined as a function of time via the unitary transformation $U(-t)$ on $\mathcal{H}$, where $U(t)$ is the evolution operator for the Hamiltonian H. They are denoted

$$|\psi; t\rangle = U(-t)|\psi\rangle$$

(2.78)

The unitary transformation redefines the observables A. They becomes time dependent and are denoted $A(t)$

$$A(t) = U(-t)AU(t)$$

(2.79)

The dynamics given by the Schrödinger equation is reabsorbed by the unitary transformation. The dynamical states are independent of time!

$$|\psi(t); t\rangle = U(-t)U(t)|\psi\rangle = |\psi\rangle$$

(2.80)

The expectation value of an observable A on a state ψ at time t is in the Heisenberg representation

$$\langle A(t) \rangle_\psi = \langle \psi(t); t | A(t) | \psi(t; , t) = \langle \psi | A(t) | \psi \rangle$$

(2.81)

The Schrödinger and Heisenberg representation are indeed equivalent, since they give the same result for the physical observable (the expectation values)

$$\langle A \rangle_{\psi(t)} = \langle A(t) \rangle_\psi$$

(2.82)

In the Heisenberg representation the Hamiltonian H remains independent of time (since it commutes with $U(t)$)

$$H(t) = H$$

(2.83)

The time evolution of the operators is given by the evolution equation

$$i\hbar \frac{d}{dt} A(t) = [A(t), H] \tag{2.84}$$

This is the quantum version of the classical Liouville equation (2.52). Of course the Schrödinger and the Heisenberg representations are the quantum analog of the two "Eulerian" and "Lagrangian" representations of classical mechanics discussed above.

For the particle in a potential the equations for Q and P are the quantum version of the classical Hamilton equations of motion

$$\frac{d}{dt} Q(t) = \frac{1}{m} P(t) \quad , \quad \frac{d}{dt} P(t) = -V'(Q(t)) \tag{2.85}$$

For an observable A which depends explicitly of time (in the Schrödinger picture), the evolution equation becomes

$$i\hbar \frac{d}{dt} A(t) = i\hbar \frac{\partial}{\partial t} A(t) + [A(t), H] \tag{2.86}$$

and taking its expectation value in some state ψ one obtains Ehrenfest theorem

$$i\hbar \frac{d}{dt} \langle A \rangle (t) = i\hbar \frac{\partial}{\partial t} \langle A \rangle (t) + \langle [A, H] \rangle (t) \tag{2.87}$$

2.3.3 Quantum Statistics and the Density Matrix

2.3.3.1 The Density Matrix

As in classical physics, in general only some partial information is available on the physical system one is interested in, or one wants to consider statistics over ensembles of states. Such situations have to be described by the concept of statistical or mixed state. But in quantum mechanics all the information one can get on a system is provided by the expectation values of its observables, since quantum mechanics contains some intrinsic indeterminism, and involves already probabilities and statistics. The pure quantum states $|\psi\rangle$ considered up to now are the quantum states which have the property that a maximal amount of information can be extracted by appropriate sets of compatible measurements on the state. The difference with classical physics is that different maximal sets of information can be extracted from the same state if one chose to perform different incompatible sets of measurements.

The mathematical concept that represents a general mixed state is the concept of density matrix. But before discussing this, one can start by noticing that, as in classical physics, an abstract statistical state ω is fully characterized by the ensemble of the expectation values $\langle A \rangle_\omega$ of all the observables A of the system, measured over the state ω.

$$\langle \mathbf{A} \rangle_\omega \quad = \quad \text{expectation value of } \mathbf{A} \text{ measured over the state } \omega \tag{2.88}$$

I denote general statistical states by Greek letters (here ω) and pure states by the bra-ket notation when there is a ambiguity. The ω here should not be confused for the notation for the symplectic form over the classical phase space of a classical system. We are dealing with quantum systems and there is no classical phase space anymore.

From the fact that the observables may be represented as an algebra $\mathcal{A}$ of operators over the Hilbert space $\mathcal{H}$, it is natural to consider that statistical states ω corresponds to linear forms over the algebra of operators $\mathcal{A}$, hence applications $\mathcal{A} \to \mathbb{C}$; $\mathbf{A} \to \langle \mathbf{A} \rangle_\omega$, with the properties

$$\langle a\mathbf{A} + b\mathbf{B} \rangle_\omega = a\langle \mathbf{A} \rangle_\omega + b\langle \mathbf{B} \rangle_\omega \qquad \text{linearity} \tag{2.89}$$

$$\langle \mathbf{A}^\dagger \rangle_\omega = \overline{\langle \mathbf{A} \rangle}_\omega \qquad \text{reality, } \bar{z} \text{ means the complex conjugate of } z \in \mathbb{C} \tag{2.90}$$

$$\langle \mathbf{A}^\dagger \mathbf{A} \rangle_\omega \geq 0 \quad \text{and} \quad \langle \mathbf{1} \rangle_\omega = 1 \qquad \text{positivity and normalization} \tag{2.91}$$

For finite dimensional Hilbert spaces and for the most common infinite dimensional cases (for physicists), any such linear a form can be represented as a normalized positive self-adjoint matrix ρ_ω

$$\rho_\omega \geq 0 \quad , \qquad \text{tr}(\rho_\omega) = 1 \tag{2.92}$$

such that for any operator $A \in \mathcal{A}$, its expectation value in the state ω is given by

$$\langle \mathbf{A} \rangle_\omega = \text{tr}(\rho_\omega \, \mathbf{A}) \tag{2.93}$$

ρ_ω is the density matrix or density operator associated to the state ω. The concept of density matrix was introduced by J. von Neumann (and independently by L. Landau and F. Bloch) in 1927. The identity (2.97) is simply the generalization of the Born rule for statistical states.

For pure states $\omega = |\psi\rangle$ the density operator is simply the rank 1 projection operator onto the state $|\psi\rangle$

$$\rho_\psi = |\psi\rangle\langle\psi| \tag{2.94}$$

One can also remark that the set of mixed states, as represented by density matrices, forms a convex set (the set of matrices satisfying (2.92)). It is convex since any statistical mixture of two mixed states ρ_1 and ρ_2 is a mixed state $\rho = p_1\rho_1 + p_2\rho_2$ ($p_1, p_2 \geq 0$ and $p_1 + p_2 = 1$). Pure states are nothing but the extremal points of the set of mixed states, i.e. the states that cannot be written as a mixture of two different states $\rho = p_1\rho_1 + p_2\rho_2$ ($\rho_1 \neq \rho_2$, $p_1, p_2 > 0$ and $p_1 + p_2 = 1$).

Before discussing some properties and features of the density matrix, let me just mention that in the physics literature, the term "state" is usually reserved to pure states, while in the mathematics literature the term "state" is used for general statistical states. The denomination "pure state" or "extremal state" is used for vectors in the Hilbert state and the associated projector. There are in fact some good mathematical reasons to use this general denomination of state.

2.3.3.2   Quantum Ensembles Versus Classical Ensembles

Let us consider a system whose Hilbert space is finite dimensional ($\dim(\mathcal{H}) = N$), in a state given by a density matrix ρ_ω. ρ_ω is a $N \times N$ self-adjoint positive matrix. It is diagonalizable and its eigenvalues are ≥ 0. If it has $1 \leq K \leq N$ orthonormal eigenvectors labeled by $|n\rangle$ ($n = 1, \cdots K$) associated with K non-zero eigenvalues p_n ($n = 1, \cdots K$) one can write

$$\rho_\omega = \sum_{n=1}^{K} p_n |n\rangle\langle n| \tag{2.95}$$

with

$$0 < p_n \leq 1, \quad \sum_n p_n = 1 \tag{2.96}$$

The expectation value of any observable $\mathbf{A}$ in the state ω is

$$\langle \mathbf{A}\rangle_\omega = \sum_n p_n \langle n|\mathbf{A}|n\rangle \tag{2.97}$$

The statistical state ω can therefore be viewed as a classical statistical mixture of the K orthonormal pure states $|n\rangle$, $n = 1, \cdots K$, the probability of the system to be in the pure state $|n\rangle$ being equal to p_n.

This point of view is useful but may be misleading. It should not be used to infer statements on how the system has been prepared. One can indeed build a statistical ensemble of independently prepared copies of the system corresponding to the state ω by picking at random, with probability p_n the system in the state $|n\rangle$. But this is not the only way to build a statistical ensemble corresponding to ω. More precisely, there are many different ways to prepare a statistical ensemble of states for the system, by picking with some probability p_α copies of the system in different states among a pre chosen set $\{|\psi_\alpha\rangle\}$ of (a priori not necessarily orthonormal) pure states, which give the same density matrix ρ_ω.

This is not a paradox. The difference between the different preparation modes is contained in the quantum correlations between the (copies of the) system and the devices used to do the preparation. These quantum correlations are fully inaccessible if one performs measurements on the system alone. The density matrix contains only the information about the statistics of the measurement on the system alone (but it encodes the maximally available information obtainable by measurements on the system only).

Another subtle point is that an ensemble of copies of a system is described by a density matrix ρ for the single system if the different copies are really independent, i.e. if there are no correlations between different copies in the ensemble, or if one simply neglect (project out) these correlations. Some apparent paradoxes arise if there are such correlations and if they must be taken into account. One must then consider the matrix density for several copies, taken as a larger composite quantum system.

2.3.3.3 The von Neumann Entropy

The "degree of uncertainty" or "lack of information" which is "contained in" a mixed quantum state ω is given by the von Neumann entropy

$$S(\omega) = -\operatorname{tr}(\rho_\omega \log \rho_\omega) = -\sum_n p_n \log p_n \tag{2.98}$$

It is the analog of the Boltzman-Gibbs entropy for a classical statistical distribution. It shares also some deep relation with Shannon entropy in information theory (more later).

The entropy of a pure state is minimal and zero. Conversely, the state of maximal entropy is the statistical state where all quantum pure states are equiprobable. It is given by a density matrix proportional to the identity, and the entropy is the logarithm of the number of accessible different (orthogonal) pure quantum state, i.e. of the dimension of the Hilbert space (in agreement with the famous Boltzmann formula $W = k_B \log N$).

$$\rho = \frac{1}{N}\mathbf{1} \quad , \quad S = \log N \quad , \quad N = \dim\mathcal{H} \tag{2.99}$$

2.3.3.4 Example: Entanglement Entropy

An important context where the density matrix has to be used is the context of open quantum systems and multipartite quantum systems. Consider a bipartite system $\mathcal{S}$ composed of two distinct subsystems $\mathcal{A}$ and $\mathcal{B}$. The Hilbert space $\mathcal{H}_\mathcal{S}$ of the pure states of $\mathcal{S}$ is the tensor product of the Hilbert space of the two subsystems

$$\mathcal{H}_\mathcal{S} = \mathcal{H}_\mathcal{A} \otimes \mathcal{H}_\mathcal{B} \tag{2.100}$$

Let us assume that the total system is in a statistical state given by a density matrix ρ_S, but that one is interested only in the subsystem $\mathcal{A}$ (or $\mathcal{B}$). In particular one can only perform easement on observables relative to $\mathcal{A}$ (or $\mathcal{B}$). Then all the information on $\mathcal{A}$ is contained in the reduced density matrix ρ_A; obtained by taking the partial trace of the density matrix for the whole system ρ_S over the (matrix indices relative to the) system $\mathcal{B}$.

$$\rho_A = \text{tr}_B [\rho_S] \tag{2.101}$$

This is simply the quantum analog of taking the marginal of a probability distribution $p(x, y)$ with respect to one of the random variables $\rho_x(x) = \int dy \, \rho(x, y)$).

If the system $\mathcal{S}$ is in a pure state $|\psi\rangle$, but if this state is entangled between $\mathcal{A}$ and $\mathcal{B}$, the reduced density matrix ρ_A is that of a mixed state, and its entropy is $S_A(\rho_A) > 0$. Indeed when considering $\mathcal{A}$ only the quantum correlations between $\mathcal{A}$ and $\mathcal{B}$ have been lost. If $\mathcal{S}$ is in a pure state the entropies $S_A(\rho_A) = S_B(\rho_B)$. This entropy is then called the entanglement entropy. Let us just recall that this is precisely one of the context where the concept of von Neumann entropy was introduced around 1927. More properties of features of quantum entropies will be given later.

2.3.3.5 Thermal States

A standard example of density matrix is provided by considering an quantum system $\mathcal{S}$ which is (weakly) coupled to a large thermostat, so that it is at equilibrium, exchanging freely energy (as well as other quantum correlations) with the thermostat, and at a finite temperature T. Then the mixed state of the system is a thermal Gibbs state (or in full generality a Kubo-Martin-Schwinger or KMS state). If the spectrum of the Hamiltonian H of the system is discrete, with the eigenstates $|n\rangle$, $n \in \mathbb{N}$ and eigenvalues (energy levels) E_n (with $E_0 < E_1 < E_2 \cdots$), the density matrix is

$$\rho_\beta = \frac{1}{Z(\beta)} \exp(-\beta H) \tag{2.102}$$

with $Z(\beta)$ the partition function

$$Z(\beta) = \text{tr} [\exp(-\beta H)] \tag{2.103}$$

and

$$\beta = \frac{1}{k_B T} \tag{2.104}$$

In the energy eigenstates basis the density matrix reads

$$\rho_\beta = \sum_n p_n |n\rangle \langle n| \tag{2.105}$$

with p_n the standard Gibbs probability

$$p_n = \frac{1}{Z(\beta)} \exp(-\beta E_n) \quad ; \quad Z(\beta) = \sum_n \exp(-\beta E_n) \tag{2.106}$$

The expectation value of an observable A in the thermal state at temperature T is

$$\langle A \rangle_\beta = \sum_n p_n \langle n|A|n \rangle = \frac{\mathrm{tr}\,[A \,\exp(-\beta H)]}{\mathrm{tr}\,[\exp(-\beta H)]} \tag{2.107}$$

For infinite systems with an infinite number of degrees of freedom, several equilibrium macroscopic states may coexist. The density matrix formalism is not sufficient and must be replaced by the formalism of KMS states (Kubo-Martin-Schwinger). This will be discussed a bit more later in connection with superselection sectors in the algebraic formalism.

2.3.3.6   Imaginary Time Formalism

Let us come back to the simple case of a quantum non-relativistic system, whose energy spectrum is bounded below (and discrete to make things simple), but unbounded from above. The evolution operator

$$U(t) = \exp\left(\frac{t}{i\hbar} H\right) \tag{2.108}$$

considered as a function of the time t, may be extended from "physical" real time $t \in \mathbb{R}$ to complex time variable, provided that

$$\mathrm{Im}(t) \leq 0 \tag{2.109}$$

More precisely, $U(t)$ as an operator, belongs to the algebra $\mathcal{B}(\mathcal{H})$ of bounded operators on the Hilbert space $\mathcal{H}$. A bounded operator A on $\mathcal{H}$ is an operator whose L^∞ norm, defined as

$$\|A\|^2 = \sup_{\psi \in \mathcal{H}} \frac{\langle \psi|A^\dagger A|\psi \rangle}{\langle \psi|\psi \rangle} \tag{2.110}$$

is finite. This is clear in the simple case where

$$U(t) = \sum_n \exp\left(\frac{t}{i\hbar}E_n\right)|n\rangle\langle n| \quad , \quad \|U(t)\| = \begin{cases} \exp\left(\frac{\text{Im}(t)}{\hbar}E_0\right) & \text{if } \text{Im}(t) \leq 0, \\ +\infty & \text{otherwise.} \end{cases}$$

$$(2.111)$$

The properties of the algebras of bounded operators and of their norm will be discussed in more details in the next section on the algebraic formulation of quantum mechanics (Fig. 2.5).

Consider now the case where t is purely imaginary

$$t = -i\tau \quad , \quad \tau > 0 \quad \text{real} \qquad U(-i\tau) = \exp\left(-\frac{\tau}{\hbar}H\right) \qquad (2.112)$$

The evolution operator has the same form than the density matrix for the system in a Gibbs state at temperature T

$$\rho_\beta = \frac{1}{Z(\beta)}U(-i\tau) \quad , \quad \beta = \frac{1}{k_B T} = \frac{\tau}{\hbar} = i\frac{t}{\hbar} \qquad (2.113)$$

There is deep analogy

$$\text{imaginary time} \sim \text{finite temperature}$$

Moreover, when considering relativistic quantum field theories in a Lorentzian metric $ds^2 = -dt^2 + d\vec{x}^2$, considering the theory at imaginary time $t = -i\tau$ implies that this imaginary time τ becomes an "Euclidean coordinate" $\tau = x^0$, and Minkowski space time becomes Euclidean space, with metric $ds^2 = d\tau^2 + d\vec{x}^2$.

These seemingly formal analogies are in fact quite important and have numerous applications. They are at the basis of Euclidean Field Theory and of the many applications of quantum field theory to statistical physics, condensed matter and

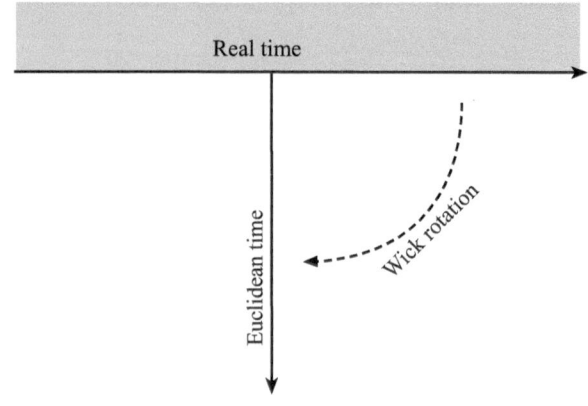

Fig. 2.5 Real time t and imaginary (Euclidean) time $\tau = it$: Wick rotation

probabilities. Reciprocally, statistical physics methods have found many applications in quantum physics and high energy physics (for instance lattice gauge theories). Considering quantum theory for imaginary time is also very useful in high energy physics, and in quantum gravity. Finally this relation between Gibbs (KMS) states and the unitary evolution operator extends in mathematics to a more general relation between states and automorphisms of operator algebras (the Tomita-Takesaki theory), that we shall discuss (very superficially) in the next chapter.

2.4 Path and Functional Integrals Formulations

2.4.1 Path Integrals

2.4.1.1 Path Integral in Configuration Space

It is known since Feynman that a very useful and efficient, if usually not yet mathematically rigorous, way to represent matrix elements of the evolution operator of a quantum system (the transition amplitudes, or "propagators") is provided by path integrals (for non-relativistic systems with a few degrees of freedom) and functional integrals (for relativistic or non relativistic systems with continuous degrees of freedoms, i.e. quantum fields).

Standard references on path integral methods on quantum mechanics and quantum field theory are the original book by Feynman and Hibbs [60], and the books by J. Zinn-Justin [113, 114].

For a single particle in an external potential this probability amplitude K for propagation from q_i at time t_i to q_f at time t_f

$$\langle q_f | U(t_f - t_i) | q_i \rangle = \langle q_f, t_f | q_i, t_i \rangle \qquad U(t) = \exp\left(\frac{t}{i\hbar} H\right) \qquad (2.114)$$

(the first notation refers to the Schrödinger picture, the second one to the Heisenberg picture) can be written as a sum of histories (or path) $q = \{q(t); t_i \le t \le t_f\}$ starting from q_i at time t_i and ending at q_f at time t_f

$$\int_{\substack{q(t_i)=q_i \\ q(t_f)=q_f}} \mathcal{D}[q] \exp\left(\frac{i}{\hbar} S[q]\right) \qquad (2.115)$$

where $S[q]$ is the classical action of the trajectory (history) (Fig. 2.6).

The precise derivation of this formula, as well as its proper mathematical definition, is obtained by decomposing the evolution of the system in a large number N of evolutions during elementary time step $\Delta t = \epsilon = t/N$, at arbitrary intermediate positions $q(t_n = n\epsilon)$, $n \in \{1, \cdots, N-1\}$, using the superposition

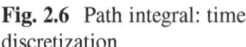

Fig. 2.6 Path integral: time discretization

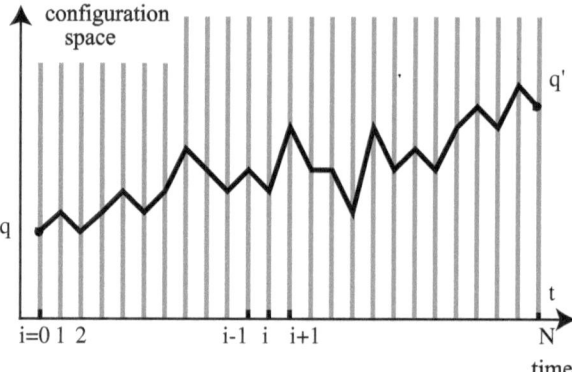

principle. One then uses the explicit formula for the propagation kernel at small time (the potential $V(q)$ may be considered as constant locally)

$$K(q_f, \epsilon, q_i, 0) \simeq \left(\frac{2i\pi\hbar\epsilon}{m}\right)^{-1/2} \exp\left(\frac{i}{\hbar}\left(\frac{m}{2}\frac{(q_f - q_i)^2}{\epsilon} - \epsilon V\left(\frac{q_f + q_i}{2}\right)\right)\right)$$
(2.116)

and one then takes the continuous time limit $\epsilon \to 0$. The precise definition of the measure over histories or paths is (from the prefactor)

$$\mathcal{D}[q] = \prod_{n=1}^{N-1}\left(dq(t_n)\left(\frac{2i\pi\hbar\epsilon}{m}\right)^{-1/2}\right)$$
(2.117)

2.4.1.2 Path Integral in Phase Space

This Lagrangian formulation of the path integral rely on a specific choice of configuration space, here the physical space of positions for the single particle. One should keep in mind that different path integrals may correspond to different quantization schemes of the same quantum theory. In particular, for this system the "Lagrangian" path integral has a "Hamiltonian" version, which corresponds to a path integral in phase space. It reads as a sum over trajectories in phase space $\{(q(t), p(t)); t_i < t \le t_f\}$

$$\int_{q(t_i)=q_i,\, q(t_f)=q_f} \mathcal{D}[q, p] \, \exp\left(\frac{i}{\hbar}\int dt\,(p\dot{q} - H(q, p)))\right)$$
(2.118)

But one must be very careful on the definition of this path integral (discretization and continuum time limit) and on the definition of the measure $\mathcal{D}[q, p]$ in order to obtain a consistent quantum theory.

2.4.2   Field Theories, Functional Integrals

The path integral representations can be generalized to relativistic quantum field theories. Let us consider the free scalar field, whose classical action (corresponding to the Klein-Gordon equation) is

$$S[\phi] = \int dt \int d^3\vec{x} \, \frac{1}{2} \left(\left(\frac{\partial \phi}{\partial t}\right)^2 - \left(\frac{\partial \phi}{\partial \vec{x}}\right)^2 - m^2 \phi^2 \right) = \int d^4 x \, \frac{1}{2} \left(-\partial^\mu \phi \, \partial_\mu \phi - m^2 \phi^2 \right)$$

(2.119)

The path integral becomes a functional integral over field configurations ϕ over space-time $\mathbf{M}^{1,3}$ of the form

$$\int \mathcal{D}[\phi] \, e^{\frac{i}{\hbar} S[\phi]} \quad , \qquad \mathcal{D}[\phi] \simeq \prod_{x \in \mathbf{M}^{1,3}} d\phi(x)$$

(2.120)

More precisely, the vacuum expectation value of time ordered product of local field operators ϕ in this quantum field theory (the so called Wightman functions, or correlation functions) can be expressed as functional integrals

$$\langle \Omega | T \phi(x_1) \cdots \phi(x_N) | \Omega \rangle = \frac{1}{Z} \int \mathcal{D}[\phi] \, e^{\frac{i}{\hbar} S[\phi]} \phi(x_1) \cdots \phi(x_N)$$

(2.121)

Z is the partition function or vacuum amplitude

$$Z = \int \mathcal{D}[\phi] \, e^{\frac{i}{\hbar} S[\phi]}$$

(2.122)

The factor Z is a normalization factor for the functional integral, so that the vacuum to vacuum transition amplitude is

$$\langle \Omega | \Omega \rangle = 1$$

The functional integral quantization method requires much care to be defined properly with full mathematical rigor. In particular the high energy-momenta / short distance singularities of quantum fields require the theory of renormalization to construct the functional integral and to check if indeed a continuum relativistic theory can be obtained and make sense as a quantum theory. This is known to be the case in some cases only (QFT in dimensions $D < 4$, some theories in $D = 4$ with an infra-red cut-off, i.e. in a finite volume).

The path integral and functional integral formulations are nevertheless invaluable tools to formulate many quantum systems and quantum field theories, grasp some of their perturbative and non-perturbative features, and perform explicit calculations. They give in particular a very simple and intuitive picture of the semiclassical regimes. they explain why the laws of classical physics can be formulated via variational principles, since classical trajectories are just the stationary phase trajectories (saddle points) dominating the sum over trajectories in the classical limit $\hbar \to 0$. In many cases they allow to treat and visualize quantum interference effects when a few semi-classical trajectories dominates (for instance for trace formulas).

Functional integral methods are also very important conceptually for quantum field theory: from the renormalization of QED to the quantization and proof of renormalisability of non abelian gauge theories, the treatment of topological effects and anomalies in QFT, the formulation of the Wilsonian renormalization group, the applications of QFT methods to statistical mechanics, etc. They thus provides a very useful way to quantize a theory, at least in semiclassical regime where one expect that the quantum theory is not in a too strong coupling regime and where quantum correlations and interference effects (possibly between non-trivial topological sectors) can be kept under control.

I shall not elaborate further here. When discussing the quantum formalism, one should keep in mind that the path integrals and functional integrals represent a very useful and powerful (if usually not fully mathematically rigorous) way to visualize, manipulate and compute transition amplitudes, i.e. matrix elements of operators. Thus functional integrals rather represent an application of the standard canonical formalism, allowing to construct the Hilbert space (or part of it) and the matrix elements of operators of a quantum theory out of the classical theory via a relatively quick and efficient recipe.

2.5   Quantum Probabilities and Reversibility

2.5.1   Is Quantum Mechanics Reversible or Irreversible?

An important aspect of classical physics and of quantum physics is the property of reversibility. By reversibility it is meant that the general formulation of the basic physical laws must be similar under time reversal. This is often stated as:

"There is no microscopic time arrow."

This does not mean that the fundamental interactions (the specific physical laws that govern our universe) are invariant under time reversal. It is known from quantum field theory that (assuming unitarity, locality and Lorentz invariance) any theory must be invariant under CPT only, the product of charge conjugation, parity and time reversal. The reversibility statement means that the dynamics of any given state, viewed forward in time (press key $\boxed{\blacktriangleright}$), is similar to the dynamics, viewed backward in time (press key $\boxed{\blacktriangleleft}$), of some other state (not necessarily the same).

This principle of reversibility is of course also very different from the macroscopic irreversibility that we experience in everyday life (the cosmological arrow of time and the expansion of the universe, the second principle of thermodynamics, some aspects of quantum measurements, irreversible behaviours in complex systems such as the Parkinson's laws [83], etc.). I am not going to discuss the issue of the emergence of irreversibility in classical physics, this would require a whole course on dynamical systems and statistical physics. Some elementary aspects will be presented in Sect. 5.5 where quantum measurements are discussed.

In classical mechanics microscopic reversibility is an obvious consequence of the Hamiltonian formulation. In quantum mechanics things are more subtle. Indeed if the evolution of a "closed system" (with no interaction with its environment and the observer) is deterministic, unitary and reversible (in particular possible quantum correlations between the system and its "outside" are kept untouched), the measurement processes over quantum systems are known to be irreversible and non-deterministic. In particular ideal projective measurements feature (via the projection postulate) the famous phenomenon of "reduction" or 'collapse" of the wave function.

> "The measurement process is an irreversible process"

This dichotomy between these two extreme classes of evolution processes has been known and has been discussed since the birth of quantum mechanics. Is the irreversible character of measurement processes a signal of the incomplete character of quantum mechanics? Is it a strange but unavoidable feature of the quantum word? Is it a macroscopic effect not so different from the occurrence of irreversibility in the classical word? Is it related to the cosmological arrow of time or to some quantum gravity effect?

I am not going to discuss these important questions here. I shall come back to a few of them in the last part of these notes. Let me just point out that, at the level of the formalism, the concepts of probabilities and of indeterminism associated with quantum measurement do not really contradict the principle of microscopic reversibility, in a certain sense that will be illustrated on a simple example below. This is in fact known since quite a long time, see for instance the well-known '64 paper by Aharonov et al. [3]. Since this point will be very important in this presentation, especially when discussing the quantum logic formalism, let me explain it on a simple, but basic example, with the usual suspects involved in quantum measurements.

2.5.2 Reversibility of Quantum Probabilities

We consider two observers, Alice and Bob, and a single quantum system S. Each of them can measure a different observable (respectively A and B) of the quantum system S (for simplicity S is taken to have a finite number of states, i.e. its Hilbert space will be finite dimensional). We take these observations to be perfect (non demolition) test measurements, i.e. yes/no measurements, represented by some selfadjoint projectors $\mathbf{P}_A$ and $\mathbf{P}_B$ such that $\mathbf{P}_A^2 = \mathbf{P}_A$ and $\mathbf{P}_B^2 = \mathbf{P}_B$, but not necessarily commuting. The eigenvalues of these operators are 1 and 0, corresponding to the two possible outcomes 1 and 0 (or TRUE and FALSE) of the measurements of the observables A and of the observable B.

Let us consider now the two following "experimental" protocols, where Alice and Bob make successive ideal measurements on a system S, and where Alice tries to guess the result of the measurement by Bob. The two protocols correspond respectively to prediction and to retrodiction.

Protocol 1—From Alice to Bob Alice gets the system S (in a state she knows nothing about). She measures A and if she finds TRUE, then she send the system to Bob, who measures B. What is the plausibility[4] for Alice that Bob will find that B is TRUE? Let us call this the conditional probability for B to be found true, A being known to be true, and denote it $P(B \leftarrow\!\!\!\mid A)$. The arrow $\leftarrow\!\!\!\mid$ denotes the causal/time ordering between the measurement of A (by Alice) and of B (by Bob) (Fig. 2.7).

Protocol 2—From Bob to Alice Alice gets the system S from Bob, and knows nothing else about S. Bob tells her that he has measured B, but does not tell her the result of his measurement, nor how the system was prepared before he performed the measurement (he may know nothing about it, he just measured B). Then Alice measures A and (if) she finds TRUE she asks herself the following question: what is the plausibility (for her, Alice) that Bob had found that B was TRUE?[5] Let us call

Fig. 2.7 Protocol 1: Alice wants to guess what will be the result of Bob's future measurement. This defines the conditional probability $P(B \leftarrow\!\!\!\mid A)$

---

[4]In a Bayesian sense.

[5]This question makes sense if for instance, Alice has made a bet with Bob. Again, and especially for this protocol, the probability has to be taken in a Bayesian sense.

this the conditional probability for B to have been found true, A being known to be true, and denote it by $P(B \mapsto A)$. The arrow $\mapsto$ denotes the causal/time ordering between the measurement of A (by Alice) and of B (by Bob) (Fig. 2.8).

Comparing the Two Protocols If S was a classical system, and the measurements were classical measurements which do not change the state of S, then the two protocols are equivalent and the two quantities equal the standard conditional probability, given by Bayes formula.

$$S \text{ classical system} : \quad P(B \leftarrow A) = P(B \mapsto A) = P(B|A) = P(B \cap A)/P(A) \, .$$

In a non classical case, assuming that the first measurement process on the system S may influence the second one, the two conditional probabilities $P(B \leftarrow A)$ and $P(B \mapsto A)$ might very well be different. A crucial and remarkable property of quantum mechanics is that these two conditional probabilities are still equal.

It is a simple but useful calculation to check this. In the first protocol $P(B \leftarrow A)$ is given by the Born rule; if Alice finds that A is TRUE and knows nothing more, her best bet is that the state of S is given by the density matrix

$$\rho_A = \mathbf{P}_A/\mathrm{Tr}(\mathbf{P}_A)$$

(equiprobabilities on the eigenspace of P_A with eigenvalue 1, this is already a Bayesian argument, already used by von Neumann [105]). Therefore for her the probability for Bob to find that B is TRUE is

$$P(B \leftarrow A) = \mathrm{tr}(\rho_A \mathbf{P}_B).$$

Fig. 2.8 Protocol 2: Alice wants to guess what was the result of Bob's past measurement. This defines the conditional probability $P(B \mapsto A)$

In the second protocol the best guess for Alice is to assume that before Bob measures B the state of the system is given by the fully equidistributed density matrix $\rho_1 = 1/\mathrm{tr}(\mathbf{1})$ (again a Bayesian argument). In this case the probability that Bob finds that B is TRUE, then that Alice finds that A is TRUE, is

$$p_1 = \mathrm{tr}(\mathbf{P}_B)/\mathrm{tr}(\mathbf{1}) \times \mathrm{tr}(\rho_B \mathbf{P}_A) \quad \text{with} \quad \rho_B = \mathbf{P}_B/\mathrm{Tr}(\mathbf{P}_B).$$

Similarly the probability that Bob finds that B is FALSE, then that Alice finds that A is TRUE is

$$p_2 = \mathrm{tr}(\mathbf{1} - \mathbf{P}_B)/\mathrm{tr}(\mathbf{1}) \times \mathrm{tr}(\rho_{\overline{B}} \mathbf{P}_A) = (\mathrm{tr}(\mathbf{P}_A) - \mathrm{tr}(\mathbf{P}_A\mathbf{P}_B))/\mathrm{tr}(\mathbf{1})$$

where $\rho_{\overline{B}} = (\mathbf{1} - \mathbf{P}_B)/\mathrm{tr}(\mathbf{1} - \mathbf{P}_B)$. The total probability is then

$$P(B \looparrowright A) = p_1 + p_2 = \mathrm{tr}(\rho_A \mathbf{P}_B).$$

Therefore, even if A and B are not compatible observables, so that the projectors $\mathbf{P}_A$ and $\mathbf{P}_B$ do not commute, one obtains in both case the same, and standard result for quantum conditional probabilities

$$S \text{ quantum system}: \quad P(B \looparrowleft A) = P(B \looparrowright A) = \mathrm{Tr}[\mathbf{P}_A\mathbf{P}_B]/\mathrm{Tr}[\mathbf{P}_A] \qquad (2.123)$$

2.5.3 Causal Reversibility

This is the basic argument. The situation studied in [3] is more complicated. It involves the selection of some initial state, a series of measurements and the postselection of some final state, but the conclusion is the same.

This reversibility property of quantum conditional probabilities is very important and is, in my opinion, a crucial feature of quantum mechanics. In this review, I shall denote it *causal reversibility*, in order not to confuse it with time reversal invariance or with the simpler property of reversibility of unitary dynamics.

Chapter 3
The Algebraic Quantum Formalism

3.1 Introduction

3.1.1 Observables as Operators

The physical observables of a quantum system are represented by the symmetric (self-adjoint) operators on the Hilbert space of pure states of the system (see Sects. 2.3.1 and in 2.3.3). They thus generate (by addition and multiplication) the set of all (not necessary symmetric) operators on the Hilbert space. This set forms an associative but non-commutative complex algebra of operators.

We have also seen that the mixed states ω (density matrices) correspond to the positive normalized linear forms on this algebra of operators, that associate to a self-adjoint operator the expectation value of the corresponding observable on the state ω.

Finally in Sect. 2.3.3.6 we have already seen that it is of interest to consider the set of "bounded operators" $\mathcal{B}(\mathcal{H})$ over some Hilbert space $\mathcal{H}$. We also mentioned that in classical mechanics, the set of smooth functions (i.e. the set of classical observables) over the phase space of a classical system form in general a Poisson algebra, i.e. a commutative algebra equipped with a Poisson bracket.

In this section I shall present and discuss the "algebraic approach" of quantum mechanics. This formulation of the principles of quantum mechanics relies precisely on the mathematical theory of algebras of operators, and is a formalisation of the above considerations. It can be viewed as an extension and as a mathematically rigorous formulation of the "canonical formalism". Of course the idea of "non-commutative observables" goes back to the "matrix mechanics" initiated by Heisenberg and was a crucial element in the elaboration the canonical formalism itself. As we shall see, in the algebraic formulation one focuses on the abstract structure of the set of observables and of the set of states of a system, and on the rules that must satisfy the probabilities associated to measurements of observables over states. The explicit realization of these (observables, states and probabilities)

© Springer International Publishing Switzerland 2015
F. David, *The Formalisms of Quantum Mechanics*, Lecture Notes in Physics 893,
DOI 10.1007/978-3-319-10539-0_3

as an algebra of operators acting on an Hilbert space representing the pure states of the system, as well as the explicit form of the probabilities as given by the Born rule, can be deduced from some more abstract, but still mostly physical axioms.

3.1.2 Operator Algebras

Let me first recall briefly which kind of operator algebras play a role in quantum mechanics. The Hilbert space $\mathcal{H}$ is a complex vector state, embodied with a scalar product (a sesquilinear form) $(\psi, \phi) \rightarrow \langle \psi | \phi \rangle = \psi \cdot \phi$ which is linear in ϕ and antilinear on ψ, symmetric, positive and non degenerate. This defines the standard norm on $\mathcal{H}$, $||\psi||^2 = \psi \cdot \psi$. To be a Hilbert space, the vector space $\mathcal{H}$ has to be complete under this norm, namely any Cauchy sequence of ψ_n, $n \in \mathbb{N}$ has a limit. Apart from the finite dimensional Hilbert spaces, the simplest and most useful Hilbert space is the separable Hilbert space, that admits a denumerable orthonormal basis (i.e. a complete basis of orthonormal vectors labelled by the integers e_n, $n \in \mathbb{N}$, $\langle e_n | e_m \rangle = \delta_{n,m}$).

Among the linear operators acting on the Hilbert space $\mathcal{H}$, an interesting class is the algebra of bounded operators $\mathcal{B}(\mathcal{H})$. An operator $A \in GL(\mathcal{H})$ is bounded if its sup-norm is finite. The norm of an operator is defined as

$$||A||^2 = \sup_{\psi \in \mathcal{H}^*} \frac{\langle A \psi | A \psi \rangle}{\langle \psi | \psi \rangle} < \infty \tag{3.1}$$

In quantum mechanics many physical observables, starting with the position operator, the momentum operator and the energy operator (the Hamiltonian), as well as the number of particles operator in quantum field theories, are not bounded operators. They may however be reconstructed from the bounded operators, within the theory of rigged Hilbert spaces (see for instance [17]).

The norm of an operator A corresponds to the diameter of its spectrum. The spectrum of A, spec(A), is the set of $z \in \mathbb{C}$ such that the operator $A - z$ is not invertible, this is the infinite dimensional generalization of the set of eigenvalues of a matrix. So the norm $||A||$ is roughly speaking the sup of the modulus of the eigenvalues of A.

This norm has many interesting property. Firstly it is indeed a norm, such that

$$||\lambda_1 A_1 + \lambda_2 A_2|| \leq |\lambda_1| \, ||A_1|| + |\lambda_2| \, ||A_2|| \tag{3.2}$$

and $\mathcal{B}(\mathcal{H})$ is complete under this norm. Moreover for products of operators it satisfy the inequality

$$||AB|| \leq ||A|| \, ||B|| \tag{3.3}$$

that makes this algebra of operators a Banach algebra. But it satisfies also the non-trivial identity ($A^\dagger$ is the adjoint of A)

$$||A^\dagger A|| = ||A||^2 = ||A^\dagger||^2 \tag{3.4}$$

that makes it a C^*-algebra. C^*-algebras have many interesting properties and are at the basis of the mathematical theory of operator algebras. They were introduced by Gelfand (under a different name). Note here that, although $\mathcal{B}(\mathcal{H})$ is a complex algebra, the "C" in the denomination is not for "complex" but rather for "compact" (the unit sphere for the norm is compact) and the "*" is for the adjoint conjugation †, which is rather denoted $*$ in the mathematical literature.

For a finite dimensional Hilbert state $\mathcal{H} = \mathbb{C}^n$, the corresponding complex C^*-algebra is the standard algebra of $n \times n$ complex matrices $M(n, \mathbb{C})$. Reciprocally, any simple finite dimensional complex C^*-algebra can be represented as a $M(n, \mathbb{C})$ matrix algebra, and this representation is unique. The situation is more interesting (both mathematically and also for physics) in the infinite dimensional case. There exists C^*-algebras (i.e. subalgebras $\mathcal{A} \subset \mathcal{B}(\mathcal{H})$ that satisfy the above conditions) that cannot be represented as the algebra of bounded operators of some "smaller" Hilbert space $\mathcal{H}'$. Moreover such C^*-algebras may have several inequivalent irreducible representations over a Hilbert space. A simple example will be given with application to one dimensional quantum mechanics later.

The mathematical theory of operator algebras was initiated by F.J. Murray and J. von Neumann in the end of the 1930's. One of J. von Neumann's motivations was precisely to formulate more precisely the mathematics of quantum mechanics, in cases where the canonical formalism and the concept of "wave function" is not sufficient or well defined. A very interesting and useful subclass of C^*-algebras is the class of the so-called W^*-algebras or von Neumann algebras. They are very important both in mathematics and for the mathematical proper formulation of quantum field theories. They will be presented and discussed (a little bit) in Sect. 3.7.

Finally a very important property of these operator algebra will be used in this chapter. Although C^*-algebras are usually defined and studied as algebras of operators acting on a Hilbert space $\mathcal{H}$, they can be also defined in an abstract way, without reference to an underlying Hilbert space. In this approach, a C^*-algebra is an abstract associative complex algebra $\mathcal{A}$, together with a conjugation $*$ (acting as the standard conjugation $A \to A^\dagger$ for operators, and a norm $||\cdot||$ satisfying the same properties of the sup-norm that I discussed above. So

$$\text{abstract } C^*\text{-algebra} = (\mathcal{A}, *, ||\cdot||) \tag{3.5}$$

The two approach are equivalent, a famous mathematical result—the GNS construction—shows that one can reconstruct all the representations of the algebra as algebras of operators acting on a Hilbert space from its abstract definition and from its states (in the sense of quantum mixed states, i.e. positive linear normalized forms) that can be constructed on the abstract algebra. Thus the Hilbert space of pure states of a quantum system can be "reconstructed" from the observables of the system.

Standard references on operator algebras in the mathematical literature are the books by J. Dixmier [39], by Sakai [90], and by P. de la Harpe and V. Jones [35].

3.1.3 The Algebraic Approach

Let me now present the algebraic formulation and what I am going to do in this chapter. In the algebraic formulation, quantum mechanics is still constructed from the classical concepts of observables and states, but one makes the assumption that the observables are not commuting objects. They will generate an associative but non-commutative algebra. The properties of this algebra of observables, and the dynamics allowed by the theory, turn out to be quite constrained, in particular by enforcing causality, locality and unitarity in order to obtain physical theories consistent with special relativity (quantum field theories). The adequate mathematical object to treat quantum field theories in a mathematically consistent way is indeed the theory of algebras of operators, in particular C* and W*-algebras, and of their representations.

As we already explained, the mathematical theory of operator algebras started in the 1930s and was developed in the 1940s and 1950s, up to now. It is still a major and expanding field of mathematics. Its applications for quantum physics, in particular for quantum field theory and quantum statistical mechanics, were notably initiated by Segal in the end of the 1940s [94] and further developed by the creation of axiomatic quantum field theory (the Wightman's axioms) [97] and shortly after through the development of algebraic quantum field theory [56] (the Haag-Kastler-Ruelle theory) in the 1960s.

The standard and excellent reference on the algebraic and axiomatic approaches to quantum field theory is the book by R. Haag, Local Quantum Physics, especially the second edition (1996) [55]. Another good, but older, reference is the book by N.N. Bogoliubov, A.A. Logunov, A.I. Oksak and I.T. Todorov [17]. Another useful reference on axiomatic QFT is the famous book by R.F. Streater and A. S. Wightman [97]. Besides the mathematical references on operator algebras given above, some mathematical references more oriented towards mathematical physics and theoretical physics are the books by Bratteli and Robinson [19] (for statistical physics) and the books by A. Connes [28] and A. Connes and M. Marcolli [29] (for high energy physics, quantum gravity and string theory).

In this chapter my goal is to give a brief and heuristic presentation of the algebraic formulation of quantum theory. So I shall try to introduce more or less precisely the mathematical concepts, but with no attempts at mathematical rigor and precision in the derivations. In contrast with the usual (and useful) mathematical presentations, where the axioms and the principles are first stated and then discussed and applied,

I shall try to motivate these axioms by logical and physical considerations, and "reconstruct" the algebraic formalism step by step, trying to make precise at which steps the different physical principles that we expect/assume for a sensible physical theory enter in the game. Of peculiar importance are the principles of causality, of reversibility and of locality (both to ensure relativistic invariance and to enforce the fact that one can causally separate domains in space-time and decompose an extended physical system into its subparts).

Such a presentation is somehow original, and I hope that it will be useful. A particular feature of this approach is that I start from the concept of observables and states of a physical system, and try to reconstruct and justify the mathematical structure (why an algebra? why a conjugation? why a sup-norm that makes the set generated by the physical observables a C^*-algebra? etc.). It then appears at that first stage that the natural structure that emerges is the structure of real C^*-algebras (i.e. algebras build on the field of real numbers $\mathbb{R}$, not necessarily on the field of complex numbers $\mathbb{C}$). Fortunately there is a mathematical theory of real C^*-algebras (less developed than the theory of complex C^*-algebras since the former ones are less interesting, and for many problems equivalent to the latter ones), that can be used as well. The only good reference on real C^*-algebras I am aware of is the monograph by Goodearl [52], and I shall refer to it.

It is only at the second stage that I shall explain which physical requirements (basically these are the requirements of locality and of causal separability) enforce the use of complex algebras and of the standard complex Hilbert space formalism.

3.2   The Algebra of Observables

3.2.1   The Mathematical Principles

A quantum system is described by its observables, its states and a causal involution acting on the observables and enforcing constraints on the states. Rather than discussing the physics concepts behind these terms, let me first give the mathematical axioms and motivate them physically after.

3.2.1.1   Observables

The physical observables of the system generate a *real associative unital algebra* $\mathcal{A}$ (whose elements will still be denoted "observables") . $\mathcal{A}$ is a linear vector space

$$\mathbf{a}, \mathbf{b} \in \mathcal{A} \quad \lambda, \mu \in \mathbb{R} \quad \lambda\mathbf{a} + \mu\mathbf{b} \in A$$

with an associative product (distributive w.r.t the addition)

$$\mathbf{a}, \mathbf{b}, \mathbf{c} \in \mathcal{A} \qquad \mathbf{ab} \in \mathcal{A} \qquad (\mathbf{ab})\mathbf{c} = \mathbf{a}(\mathbf{bc}) \tag{3.6}$$

and a multiplicative unity element **1** such that

$$\mathbf{1a} = \mathbf{a1} = \mathbf{a}, \quad \forall\, \mathbf{a} \in \mathcal{A} \tag{3.7}$$

I shall make precise later what is meant by "physical observables".

3.2.1.2 The *-Conjugation

There is an involution * on $\mathcal{A}$ (denoted conjugation). It is an anti-automorphism whose square is the identity. This means that

$$(\lambda \mathbf{a} + \mu \mathbf{b})^\star = \lambda \mathbf{a}^\star + \mu \mathbf{b}^\star \tag{3.8}$$

and

$$(\mathbf{a}^\star)^\star = \mathbf{a} \qquad (\mathbf{ab})^\star = \mathbf{b}^\star \mathbf{a}^\star \tag{3.9}$$

3.2.1.3 States

Each state φ associates to an observable $\mathbf{a}$ its expectation value $\varphi(\mathbf{a}) \in \mathbb{R}$ in the state φ. The states satisfy

$$\varphi(\lambda \mathbf{a} + \mu \mathbf{b}) = \lambda \varphi(\mathbf{a}) + \mu \varphi(\mathbf{b}) \tag{3.10}$$

and

$$\varphi(\mathbf{a}^\star) = \varphi(\mathbf{a}) \qquad \varphi(\mathbf{1}) = 1 \qquad \varphi(\mathbf{a}^\star \mathbf{a}) \geq 0 \tag{3.11}$$

The set of states is denoted $\mathcal{E}$. It is natural to assume that it allows to discriminate between observables, i.e.

$$\forall\, \mathbf{a} \neq \mathbf{b} \in \mathcal{A} \text{ (and } \neq \mathbf{0}), \exists\, \varphi \in \mathcal{E} \text{ such that } \varphi(\mathbf{a}) \neq \varphi(\mathbf{b}) \tag{3.12}$$

For the unfamiliar reader the symbol $\forall$ means "for all" and $\exists$ means "there exists".

I do not discuss the concepts of time and dynamics at that stage. This will be done later. I first discuss the relation between these "axioms" and the physical concepts of causality, reversibility and probabilities.

3.2.2 Physical Discussion

3.2.2.1 Observables and Causality

In quantum physics, the concept of physical observable corresponds both to an operation on the system (measurement) and to the response of the system (result on the measure), but I shall not elaborate further. I already discussed why in classical physics observables form a real commutative algebra. The removal of the commutativity assumption is the simplest modification imaginable compatible with the uncertainty principle (Heisenberg 1925).

Keeping the mathematical structure of an associative but non commutative algebra reflects the assumption that there is still some concept of "causal ordering" between observables (not necessarily physical), in a formal but loose sense. Indeed the multiplication and its associativity means that we can "combine" successive observables, e.g. **ab** $\simeq$ (**b** then **a**), in a linear process such that ((**c** then **b**) then **a**) $\simeq$ (**c** then (**b** then **a**)). This "combination" is different from the concept of "successive measurement".

Without commutativity the existence of an addition law is already a non trivial fact, it means that we can "combine" two non compatible observations into a new one whose mean value is always the sum of the first two mean values.

These operations of addition and multiplication of observables are in fact more natural in the context of relativistic theories, via the analyticity properties of correlation functions and the short time and short distance expansions.

3.2.2.2 The *-Conjugation and Reversibility

The existence of the involution (or conjugation) * is the second and very important feature of quantum physics. It implies that although the observables do not commute, there is no favored arrow of time (or causal ordering) in the formulation of a physical theory. To any causal description of a system in term of a set of observables {**a**, **b**, ...} corresponds an equivalent "anti-causal" description it terms of conjugate observables {**a***, **b***, ...}. In other word the properties of the * conjugation amount to assuming microscopic reversibility. Again although there is no precise concept of time or dynamics yet, the involution * must not be confused with the time reversal operator **T** (which may or may not be a symmetry of the dynamics).

3.2.2.3 States, Measurements and Probabilities

The states φ are the simple generalisation of the classical concept of statistic (or probabilistic) states describing our knowledge of a system through the expectation value of the outcome of measurements for each possible observables. At that stage

I do not assume anything about whether there are states such that all the values of the observables can be determined or not. Thus a state can be viewed also as the characterization of all the information which can be extracted from a system through a measurement process (this is the point of view often taken in quantum information theory). I do not consider how states are prepared, nor how the measurements are performed (this is the object of the subpart of quantum theory known as the theory of quantum measurement) and just look at the consistency requirements on the outcome of measurements.

The "expectation value" $\varphi(\mathbf{a})$ of an observable $\mathbf{a}$ has not been defined precisely either at that stage. In statistics the expectation of an observable can be considered as well as given by the average of the outcome of measurements $\mathbf{a}$ over many realisations of the system in the same state (frequentist point of view) or as the sum over the possible outcomes a_i times the plausibility for the outcomes in a given state (Bayesian point of view). As already done in Sect. 2.5.2, and we shall see in the discussion of the algebraic formalism and of the quantum logic formalism, discussing the role of reversibility amounts to treat simultaneously the statistics for predictions (which can be done using the frequentist point of view) and the statistics for retrodictions (which is better done using the Bayesian point of view). Therefore in my opinion both point of views have to be considered on a same footing, and are somehow unified, in the quantum formalism.

The linearity of the states considered as function over the observables Eq. (3.10) follows from (or is equivalent to) the assumptions that the observables form a linear vector space on $\mathbb{R}$. The very important condition in (3.11)

$$\varphi(\mathbf{a}^*) = \overline{\varphi(\mathbf{a})}$$

for any $\mathbf{a}$ follows from the assumption of reversibility discussed above. If it was not satisfied, there would be observables that would allow to favor one causal ordering, irrespective of the possible dynamics and of the possible states of the system.

The positivity condition $\varphi(\mathbf{a}^*\mathbf{a}) \geq 0$ ensures that the states have a probabilistic interpretation, so that on any state the expectation value of a positive observable is positive, and that there are no negative probabilities, in other word it will ensure unitarity. It is the simplest consistent positivity condition compatible with reversibility, and in fact the only possible without assuming more structure on the observables. Of course the condition $\varphi(\mathbf{1}) = 1$ is the normalisation condition for probabilities.

3.2.3   Physical Observables and Pure States

Three important concepts follow from the mathematical principles assumed in Sect. 3.2.1, and tentatively justified physically in Sect. 3.2.2.

3.2.3.1 Physical (Symmetric) Observables

An observable $\mathbf{a} \in \mathcal{A}$ is denoted symmetric (self adjoint, or self-conjugate) if $\mathbf{a}^* = \mathbf{a}$. Symmetric observables correspond to the physical observables, which are actually measurable. Observable such that $\mathbf{a}^* = -\mathbf{a}$ are denoted skew-symmetric (anti-symmetric or anti-conjugate). They do not correspond to physical observables since for such observables $\varphi(a) = 0$ but they must be included in the formalism in order to have a consistent algebraic structure.

3.2.3.2 Pure States

The set of states $\mathcal{E}$ is a convex subset of the set of real linear forms on $\mathcal{A}$ (the dual of $\mathcal{A}$). Indeed if φ_1 et φ_2 are two states and $0 \le x \le 1$, $\varphi = x\varphi_1 + (1 - x)\varphi_2$ is also a state. This corresponds to the fact that any statistical mixture of two statistical mixtures is a statistical mixture. The extremal points in $\mathcal{E}$, i.e. the states which cannot be written as a statistical mixture of two different states in $\mathcal{E}$, are called the pure states. Non pure states are called mixed states. If a system is in a pure state one cannot get more information from this system than the information that we have already.

3.2.3.3 Bounded Observables

I just need to impose two additional technical and natural assumptions: (1) for any observable $\mathbf{a} \ne 0$, there is a state φ such that $\varphi(\mathbf{a}^*\mathbf{a}) > 0$, if this is not the case, the observable $\mathbf{a}$ is indistinguishable from the observable 0 (which is always false); (2) $\sup_{\varphi \in \mathcal{E}} \varphi(\mathbf{a}^*\mathbf{a}) < \infty$, i.e. I restrict $\mathcal{A}$ to the algebra of bounded observables, this will be enough to characterize the system.

3.3 The C*-Algebra of Observables

The involution * et the existence of the states $\varphi \in \mathcal{E}$ on $\mathcal{A}$ strongly constrain the structure of the algebra of observables and of its representations. Indeed this allows to associate to $\mathcal{A}$ a unique norm $\| \cdot \|$ with some specific properties. This norm makes $\mathcal{A}$ a C*-algebra, and more precisely a real abstract C*-algebra. This structure justifies the standard representation of quantum mechanics where pure states are elements of an Hilbert space and physical observables are self-adjoint operators.

3.3.1   The Norm on Observables, $\mathcal{A}$ is a Banach Algebra

Let us consider the function $\mathbf{a} \to ||\mathbf{a}||$ from $\mathcal{A} \to \mathbb{R}^+$ defined by

$$||\mathbf{a}||^2 = \sup_{\text{states } \varphi \in \mathcal{E}} \varphi(\mathbf{a}^{\star}\mathbf{a}) \tag{3.13}$$

We have assumed that $||\mathbf{a}|| < \infty$, $\forall a \in \mathcal{A}$ and that $||\mathbf{a}|| = 0 \iff \mathbf{a} = 0$ (this is equivalent to $\mathbf{a} \neq 0 \implies \exists \varphi \in \mathcal{E}$ such that $\varphi(\mathbf{a}^{\star}\mathbf{a}) \neq 0$). It is easy to show that $||\cdot||$ is a norm on $\mathcal{A}$, such that

$$||\lambda\mathbf{a}|| = |\lambda|\,||\mathbf{a}||\quad,\qquad ||\mathbf{a}+\mathbf{b}|| \leq ||\mathbf{a}|| + ||\mathbf{b}||\quad,\qquad ||\mathbf{ab}|| \leq ||\mathbf{a}||\,||\mathbf{b}|| \tag{3.14}$$

If $\mathcal{A}$ is not closed for this norm, we can take its completion $\overline{\mathcal{A}}$. The algebra of observables is therefore a real Banach algebra.

Derivation The first identity in (3.14) comes from the definition and the linearity of states. Taking $\mathbf{c} = x\mathbf{a} + (1-x)\mathbf{b}$ and using the positivity of $\varphi(c^*c) \geq 0$ for any $x \in \mathbb{R}$ one obtains Schwartz inequality $\varphi(a^*b)^2 = \varphi(a^*b)\varphi(b^*a) \leq \varphi(a^*a)\varphi(b^*b)$, $\forall\, a, b \in \mathbf{A}$. This implies the second inequality. The third inequality comes from the fact that if $\varphi \in \mathcal{E}$ and $b \in \mathcal{A}$ are such that $\varphi(b^*b) > 0$, then φ_b defined by $\varphi_b(a) = \frac{\varphi(b^*ab)}{\varphi(b^*b)}$ is also a state for $\mathcal{A}$. Then $||ab||^2 = \sup_\varphi \varphi(b^*a^*ab) = \sup_\varphi (\varphi_b(a^*a)\varphi(b^*b)) \leq \sup_\psi \psi(a^*a) \cdot \sup_\varphi \varphi(b^*b) = ||a||^2\,||b||^2$.

3.3.2   The Observables form a Real C*-Algebra

Moreover the norm satisfies the two non-trivial properties.

$$||\mathbf{a}^*\mathbf{a}|| = ||\mathbf{a}||^2 = ||\mathbf{a}^*||^2 \tag{3.15}$$

and

$$\mathbf{1} + \mathbf{a}^*\mathbf{a}\quad\text{is invertible}\quad\forall\,\mathbf{a} \in \mathcal{A} \tag{3.16}$$

These two properties are equivalent to state that $\mathcal{A}$ is a real C*-algebra. The first condition (3.15) is sometimes called the C* condition. It has already been discussed for complex algebras. The second condition (3.16) is specific to real algebras. For a discussion of the definition of real C*-algebras and of their properties, that will be used below, I refer to the review by Goodearl [52].

Derivation One has $||a^*a|| \leq ||a||\,||a^*||$. Schwartz inequality implies that $\varphi(a^*a)^2 \leq \varphi\big((a^*a)^2\big)\varphi(1)$, hence $||a||^2 \leq ||a^*a||$. This implies (3.15).

To obtain (3.16), notice that if $1 + a^*a$ is not inversible, there is a $b \neq 0$ such that $(1 + a^*a)b = 0$, hence $b^*b + (ab)^*(ab) = 0$. Since there is a state φ such that $\varphi(b^*b) \neq 0$, either $\varphi(b^*b) < 0$ or $\varphi((ab)^*(ab)) < 0$, this contradicts the positivity of states.

The full consequences of the fact that $\mathcal{A}$ is a real C*-algebra will be discussed in next subsection. Before this let me introduce first the concept of spectrum of an observable.

3.3.3 Spectrum of Observables and Results of Measurements

Here I discuss in a slightly more precise way the relationship between the spectrum of observables and results of measurements. The spectrum[1] of an element $\mathbf{a} \in \mathcal{A}$ is defined as

$$\mathrm{Sp}^c(\mathbf{a}) = \{ z \in \mathbb{C} : (z - \mathbf{a}) \text{ not inversible in the complexified algebra } \mathcal{A}_{\mathbb{C}} \text{ of } \mathcal{A} \} \ .$$

The spectral radius of $\mathbf{a}$ is defined as

$$r^c(\mathbf{a}) = \sup(|z|; \ z \in \mathrm{Sp}^c(\mathbf{a}))$$

For a real C*-algebra it is known that the norm $|| \cdot ||$ defined by (3.13) is

$$||a||^2 = r^c(\mathbf{a}^*\mathbf{a})$$

Moreover the spectrum of any physical observable (symmetric) is real

$$\mathbf{a} = \mathbf{a}^* \implies \mathrm{Sp}^c(\mathbf{a}) \subset \mathbb{R}$$

and for any $\mathbf{a}$, the product $\mathbf{a}^*\mathbf{a}$ is a symmetric positive element of $\mathcal{A}$, i.e. its spectrum is real and positive

$$\mathrm{Sp}^c(\mathbf{a}^*\mathbf{a}) \subset \mathbb{R}^+$$

Finally, for any (continuous) real function $F \ \mathbb{R} \to \mathbb{R}$ and any $\mathbf{a} \in \mathcal{A}$ one can define the observable $F(\mathbf{a})$. Now consider a physical observable $\mathbf{a}$. Physically, measuring $F(\mathbf{a})$ amounts to measure $\mathbf{a}$ and when we get the real number A as the result of the measurement, return $F(A)$ as a result of the measure of $F(\mathbf{a})$ (this is fully consistent with the algebraic definition of $F(\mathbf{a})$ since $F(\mathbf{a})$ commutes with $\mathbf{a}$). Then it can be shown easily that the spectrum of $F(\mathbf{a})$ is the image by F of the spectrum of $\mathbf{a}$, i.e.

$$\mathrm{Sp}^c(F(\mathbf{a})) = F(\mathrm{Sp}^c(\mathbf{a}))$$

---

[1]The exact definition of the spectrum is slightly different for a general real Banach algebra.

In particular, assuming that the spectrum is a discrete set of points, let us choose for F the function

$$F[\mathbf{a}] = 1/(z\mathbf{1} - \mathbf{a})$$

For any state φ, the expectation value of this observable on the state φ is

$$E_\varphi(z) = \varphi(1/(z\mathbf{1} - \mathbf{a})$$

and is an analytic function of z away from the points of the spectrum $\mathrm{Sp}^c(\mathbf{a}))$. Assuming that the singularity at each z_p is a single pole, the residue of $E_\varphi(z)$ at z_p is nothing but

$$Res_{z_p} E_\varphi = \varphi(\delta(\mathbf{a} - z_p\mathbf{1}))$$

$$= \text{probability to obtain } z_p \text{ when measuring } \mathbf{a} \text{ on the state } \varphi \qquad (3.17)$$

with $\delta(z)$ the Dirac distribution.

This implies that for any physical observable $\mathbf{a}$, its spectrum is the set of all the possible real numbers z_p returned by a measurement of $\mathbf{a}$. This is one of the most important axioms of the standard formulation of quantum mechanics, and we see that it is a consequence of the axioms in this formulation. Of course the probability to get a given value z_p (an element of the spectrum) depends on the state φ of the system, and it is given by (3.17) which is nothing but the Born rule, as obtained from this abstract definition of the states.

3.3.4 Complex C*-Algebras

The theory of operator algebras (C*-algebras and W*-algebras) and their applications almost exclusively deal with complex algebras, i.e. algebras over $\mathbb{C}$. In the case of quantum physics we shall see a bit later why quantum mechanics and quantum field theories must be represented by complex C*-algebras. I give here some mathematical definitions.

Abstract complex C*-algebras and complex states ϕ are defined as in Sect. 3.2.1. A complex C*-algebra $\mathfrak{A}$ is a complex associative involutive algebra. The involution is now anti-linear

$$(\lambda\mathbf{a} + \mu\mathbf{b})^\star = \bar{\lambda}\mathbf{a}^\star + \bar{\mu}\mathbf{b}^\star \qquad \lambda, \mu \in \mathbb{C}$$

$\bar{z}$ denotes the complex conjugate of a complex number z. $\mathfrak{A}$ is equipped with a norm $\mathbf{a} \to ||\mathbf{a}||$ which still satisfy the C* condition (3.15),

$$||\mathbf{a}^*\mathbf{a}|| = ||\mathbf{a}||^2 = ||\mathbf{a}^*||^2 \qquad (3.18)$$

and it is closed under this norm. The condition (3.16) is not necessary any more (it follows from (3.18) for complex algebras).

The states are defined now as the complex linear forms ϕ on $\mathfrak{A}$ which satisfy

$$\phi(\mathbf{a}^*) = \overline{\phi(\mathbf{a})} \qquad \phi(\mathbf{1}) = 1 \qquad \phi(\mathbf{a}^*\mathbf{a}) \geq 0 \qquad (3.19)$$

Every complex C*-algebra $\mathfrak{A}$ can be considered as a real C*-algebra $\mathcal{A}_{\mathbb{R}}$ (by considering $i = \sqrt{-1}$ as an element $\mathbf{i}$ of the center of $\mathcal{A}_{\mathbb{R}}$) but the reverse is not true in general. For instance the algebra of 2×2 real matrices $M_2(\mathbb{R})$ is not a complex algebra.

However if a real algebra $\mathcal{A}_{\mathbb{R}}$ has an element (denoted $\mathbf{i}$) in its center $\mathcal{C}$ that is isomorphic to $\sqrt{-1}$, i.e. $\mathbf{i}$ is such that

$$\mathbf{i}^* = -\mathbf{i}, \quad \mathbf{i}^2 = -\mathbf{1}, \quad \mathbf{ia} = \mathbf{ai} \ \forall \ \mathbf{a} \in \mathcal{A}_{\mathbb{R}} \qquad (3.20)$$

then the algebra $\mathcal{A}_{\mathbb{R}}$ is isomorphic to a complex algebra $\mathcal{A}_{\mathbb{C}} = \mathfrak{A}$. One identifies $x\mathbf{1} + y\mathbf{i}$ with the complex scalar $z = x + iy$. The conjugation *, which is linear on $\mathcal{A}_{\mathbb{R}}$, is now anti-linear on $\mathcal{A}_{\mathbb{C}}$. One can associate to each $\mathbf{a} \in \mathcal{A}_{\mathbb{R}}$ its real and imaginary part

$$\mathrm{Re}(\mathbf{a}) = \frac{\mathbf{a} + \mathbf{a}^*}{2}, \quad \mathrm{Im}(\mathbf{a}) = \mathbf{i}\frac{\mathbf{a}^* - \mathbf{a}}{2} \qquad (3.21)$$

and write it in $\mathcal{A}_{\mathbb{C}}$

$$\mathbf{a} = \mathrm{Re}(\mathbf{a}) + i\,\mathrm{Im}(\mathbf{a}) \qquad (3.22)$$

To any real state (and in fact to any real linear form) $\varphi_{\mathbb{R}}$ on $\mathcal{H}_{\mathbb{R}}$ one associates the complex state (the complex linear form) $\phi_{\mathbb{C}}$ on $\mathcal{A}_{\mathbb{C}}$ defined as

$$\phi_{\mathbb{C}}(\mathbf{a}) = \varphi_{\mathbb{R}}(\mathrm{Re}(\mathbf{a})) + i\,\varphi_{\mathbb{R}}(\mathrm{Im}(\mathbf{a})) \qquad (3.23)$$

It has the expected properties for a complex state on the complex algebra $\mathfrak{A}$.

3.4 The GNS Construction, Operators and Hilbert Spaces

General theorems show that abstract C*-algebras can always be represented as algebra of operators on some Hilbert space. This is the main reason why pure states are always represented by vectors in a Hilbert space and observables as operators. Let me briefly consider how this works.

3.4.1 Finite Dimensional Algebra of Observables

Let me first consider the case of finite dimensional algebras, which corresponds to quantum systems with a finite number of independent quantum states. This is the case considered in general in quantum information theory.

If $\mathcal{A}$ is a finite dimensional real algebra, one can show by purely algebraic methods that $\mathcal{A}$ is a direct sum of matrix algebras over $\mathbb{R}$, $\mathbb{C}$ or $\mathbb{H}$ (the quaternions). See [52] for details. The idea is to show that the C*-algebra conditions implies that the real algebra $\mathcal{A}$ is semi-simple (it cannot have a nilpotent two-sided ideal) and to use the Artin-Wedderburn theorem. One can even relax the positivity condition on states that $\varphi(\mathbf{a}^*\mathbf{a}) \geq 0$ for all $\mathbf{a} \in \mathcal{A}$, and replace it by the weaker positivity condition that $\varphi(\mathbf{a}^2) \geq 0$ only for physical symmetric observables such that $\mathbf{a} = \mathbf{a}^*$, which is physically somewhat more satisfactory (F. David unpublished, probably known in the math literature...). This is physically more satisfactory in my opinion, since at that stage it is not completely obvious that any positive physical observable can be represented as the square of a physical observable). The conclusion is that any finite dimensional real C*-algebra is a direct sum of matrix algebras, of the form

$$\mathcal{A} = \bigoplus_i M_{n_i}(K_i) \qquad K_i = \mathbb{R}, \mathbb{C}, \mathbb{H} \tag{3.24}$$

The index i labels the components of the center of the algebra. Any observable reads

$$\mathbf{a} = \oplus_i \mathbf{a}_i , \quad \mathbf{a}_i \in \mathcal{A}_i = M_{n_i}(K_i)$$

The multiplication corresponds to the standard matrix multiplication and the involution * to the standard conjugation (transposition, transposition+complex conjugation and transposition+conjugation respectively for real, complex and quaternionic matrices). One thus recovers the familiar matrix ensembles of random matrix theory.

Any state ω can be written as

$$\omega(\mathbf{a}) = \sum_i p_i \, \mathrm{tr}(\rho_i \mathbf{a}_i) \qquad p_i \geq 0, \quad \sum_i p_i = 1$$

and the ρ_i's some symmetric positive normalised matrices in each $\mathcal{A}_i$

$$\rho_i \in \mathcal{A}_i = M_{n_i}(K_i), \quad \rho_i = \rho_i^*, \quad \mathrm{tr}(\rho_i) = 1, \quad \rho_i \geq 0$$

The algebra of observables is indeed a subalgebra of the algebra of operators on a finite dimensional real Hilbert space $\mathcal{H} = \bigoplus_i K_i^{n_i}$ ($\mathbb{C}$ and $\mathbb{H}$ being considered as two dimensional and four dimensional real vector spaces respectively). But it is not necessarily the whole algebra $\mathcal{L}(\mathcal{H})$. The system corresponds to a disjoint collection of standard quantum systems described by their Hilbert space $\mathcal{H}_i \simeq K_i^{\oplus n_i}$ and their algebra of observables $\mathcal{A}_i$. This decomposition is (with a bit of abuse of language) a decomposition into "superselection sectors".[2] The ρ_i are the quantum density matrices corresponding to the state. The p_i's correspond to the classical probability to be in a given sector, i.e. in a state described by $(\mathcal{A}_i, \mathcal{H}_i)$.

A pure state is (the projection onto a) single vector $|\psi_i\rangle$ in a single sector $\mathcal{H}_i$. Linear superpositions of pure states in different sectors $|\psi\rangle = \sum_i c_i |\psi_i\rangle$ do not make sense, since they do not belong to the representation of $\mathcal{A}$. No observable **a** in $\mathcal{A}$ allows to discriminate between the seemingly-pure-state $|\psi\rangle\langle\psi|$ and the mixed state $\sum_i |c_i|^2 |\psi_i\rangle\langle\psi_i|$. Thus the different sectors can be viewed as describing completely independent systems with no quantum correlations, in other word really parallel universes with no possible interaction or communication between them.

3.4.2 Infinite Dimensional Real Algebra of Observables

This result generalizes to the case of infinite dimensional real C*-algebras, but it is much more difficult to prove, analysis and topology enter in the game and the fact that the algebra is closed under the norm is crucial (for a physicist this is a natural requirement).

Theorem (Ingelstam NN [52, 63]) For any real C*-algebra, there exists a real Hilbert space $\mathcal{H}$ such that $\mathcal{A}$ is isomorphic to a real symmetric closed real sub-algebra of the algebra $B(\mathcal{H})$ of bounded operators on $\mathcal{H}$.

Now any real algebra of symmetric operators on a real Hilbert space $\mathcal{H}$ may be extended (by standard complexification) into a complex algebra of self-adjoint operator on a Hilbert space $\mathcal{H}_{\mathbb{C}}$ on $\mathbb{C}$ and thus one can reduce the study of real algebra to the study of complex algebra. In particular the theory of representations of real C*-algebra is not really richer than that of complex C*-algebra and mathematicians usually consider only the later case.

I will discuss later why in quantum physics one should restrict oneself also to complex algebras. But note that in physics real (and quaternionic) algebra of observables do appear as the subalgebra of observables of some system described by a complex Hilbert space, subjected to some additional symmetry constraint (time reversal invariance **T** for real algebra, time reversal and an additional SU(2) invariance for quaternionic algebras).

---

[2]See below for a more precise definition and discussion. For many authors the term of superselection sectors is reserved to infinite dimensional algebras which do have inequivalent representations.

3.4.3 The Complex Case, the GNS Construction

Let me discuss more the case of complex C*-algebras, since their representation in term of Hilbert spaces are simpler to deal with. The famous GNS construction (Gelfand-Naimark-Segal [47, 93]) allows to construct the representations of the algebra of observables in term of its pure states. It is interesting to see the basic ideas, since they allow to understand how the Hilbert space of physical pure states emerges from the abstract[3] concepts of observables and mixed states.

The idea is quite simple. To every state ϕ we associate a representation of the algebra $\mathcal{A}$ in a Hilbert space $\mathcal{H}_\phi$. This is done as follows. The state ϕ allows to define a bilinear form $\langle\ |\ \rangle_\phi$ on $\mathcal{A}$, considered as a vector space on $\mathbb{C}$, through

$$\langle\mathbf{a}|\mathbf{b}\rangle_\phi = \phi(\mathbf{a^*b}) \tag{3.25}$$

This form is positive ≥ 0 but is not strictly positive > 0, since there are in general isotropic (or null) vectors such that $\langle\mathbf{a}|\mathbf{a}\rangle_\phi = 0$. Thus $\mathcal{A}$, considered as a vector space equipped with this norm is only a pre-Hilbert space. However, thanks to the C*-condition, the set $\mathcal{I}_\phi$ of these null vectors form a linear subspace $\mathcal{I}_\phi$ of $\mathcal{A}$.

$$\mathcal{I}_\phi = \{\mathbf{a} \in \mathcal{A} : \langle\mathbf{a}|\mathbf{a}\rangle_\phi = 0\} \tag{3.26}$$

Taking the completion of the quotient space of $\mathcal{A}$ by $\mathcal{I}_\phi$ one obtains a vector space $\mathcal{H}_\phi$

$$\mathcal{H}_\phi = \overline{\mathcal{A}/\mathcal{I}_\phi} \tag{3.27}$$

When there is no ambiguity, if $\mathbf{a}$ is an element of the algebra $\mathcal{A}$ (an observable), we denote by $|a\rangle$ the corresponding vector in the Hilbert space $\mathcal{H}_\phi$, that is the equivalent class of $\mathbf{a}$ in $\mathcal{H}_\phi$

$$|a\rangle = \{\mathbf{b} \in \mathcal{A} : \mathbf{b} - \mathbf{a} \in \mathcal{I}_\phi\} \tag{3.28}$$

On this space $\mathcal{H}_\phi$ the scalar product $\langle a|b\rangle$ is > 0 (and $\mathcal{H}_\phi$ is closed) hence $\mathcal{H}_\phi$ is a Hilbert space.

The algebra $\mathcal{A}$ acts linearly on $\mathcal{H}_\phi$ through the representation π_ϕ (in the space of bounded linear operators $B(\mathcal{H}_\phi)$ on $\mathcal{H}_\phi$) defined as

$$\pi_\phi(\mathbf{a})|b\rangle = |ab\rangle \tag{3.29}$$

---

[3]in the mathematic sense: they are not defined with reference to a given representation such as operators in Hilbert space, path integrals, etc.

If one considers the vector $|\xi_\phi\rangle = |1\rangle \in \mathcal{H}_\phi$ (the equivalence class of the operator identity $\mathbf{1} \in \mathcal{A}$), it is of norm 1 and it is such that

$$\phi(\mathbf{a}) = \langle \xi_\phi | \pi_\phi(\mathbf{a}) | \xi_\phi \rangle \tag{3.30}$$

(this follows basically from the definition of the representation). Moreover this vector $|\xi_\phi\rangle$ is cyclic, this means that the action of the operators on this vector allows to recover the whole Hilbert space $\mathcal{H}_\phi$, more precisely

$$\overline{\pi_\phi(\mathcal{A})|\xi_\phi\rangle} = \mathcal{H}_\phi \tag{3.31}$$

However this representation is in general neither faithful (different observables may be represented by the same operator, i.e. the mapping π_ϕ is not injective), nor irreducible ($\mathcal{H}_\phi$ has invariant subspaces). The most important result of the GNS construction is the following theorem.

Theorem (Gelfand-Naimark 43) The representation π_ϕ is irreducible if and only if ϕ is a pure state.

The proof is standard and may be found in [35]. This theorem has far reaching consequences. First it implies that the algebra of observables $\mathcal{A}$ has always a faithful representation in some big Hilbert space $\mathcal{H}$. Any irreducible representation π of $\mathcal{A}$ in some Hilbert space $\mathcal{H}$ is unitarily equivalent to the GNS representation π_ϕ constructed from a unit vector $|\xi\rangle \in \mathcal{H}$ by considering the state

$$\phi(\mathbf{a}) = \langle \xi | \pi(\mathbf{a}) | \xi \rangle$$

Equivalent Pure States Two pure states ϕ and ψ are equivalent if their GNS representations π_ϕ and π_ψ are equivalent. Then ϕ and ψ are unitarily equivalent, i.e. there is a unitary element $\mathbf{u}$ of $\mathcal{A}$ ($\mathbf{u}^*\mathbf{u} = \mathbf{1}$) such that $\phi(\mathbf{a}) = \psi(\mathbf{u}^*\mathbf{au})$ for any $\mathbf{a}$. As a consequence, to this pure state ψ (which is unitarily equivalent to ϕ) is associated a unit vector $|\psi\rangle = \pi_\phi(\mathbf{u})|\xi_\phi\rangle$ in the Hilbert space $\mathcal{H} = \mathcal{H}_\phi$, and we have the representation

$$\psi(\mathbf{a}) = \langle \psi | A | \psi \rangle \quad , \qquad A = \pi_\phi(\mathbf{a}) \tag{3.32}$$

In other word, all pure states which are equivalent can be considered as projection operators $|\psi\rangle\langle\psi|$ on some vector $|\psi\rangle$ in the same Hilbert space $\mathcal{H}$. Any observable $\mathbf{a}$ is represented by some bounded operator A and the expectation value of this observable in the state ψ is given by the Born formula (3.32). Equivalent classes of equivalent pure states are in one to one correspondence with the irreducible representations of the algebra of observables $\mathcal{A}$.

The standard explicit formulation of quantum mechanics in terms of operators, state vectors and density matrices is thus recovered from the abstract formulation

3.5 Why Complex Algebras?

In the mathematical presentation of the formalism that I gave here, real algebras play the essential role. However it is known that quantum physics must be described by complex algebras. There are several arguments that point towards the necessity of complex algebras, besides the fact that this actually works. Indeed one must still take into account some essential physical features of the quantum word: time, dynamics and locality.

3.5.1 Dynamics

Firstly, if one wants the quantum system to have a "classical limit" corresponding to a classical Hamiltonian system, one would like to have conjugate observables P_i, Q_i whose classical limit are conjugate coordinates p_i, q_i with a correspondence between the quantum commutators and the classical Poisson brackets

$$[Q, P] \to i\{p, q\} \tag{3.33}$$

Thus anti-symmetric operators must be in one to one correspondence with symmetric ones. This is possible only if the algebra of operators is a complex one, i.e. if it contains an **i** element in its center.

Another, but related, argument goes as follows: if one wants to have a time evolution group of inner automorphism acting on the operators and the states, it is given by unitary evolution operators $U(t)$ of the form

$$U(t) = \exp(tA) \quad , \qquad A = -A^* \tag{3.34}$$

This corresponds to have an Hamiltonian dynamics with a physical observable corresponding to a conserved energy, and given by a Schrödinger equation. This is possible only if the algebra is complex, so that we can write

$$A = -iH \tag{3.35}$$

There has been various attempts to construct realistic quantum theories of particles or fields based on strictly real Hilbert spaces, most notably by Stueckelberg and his collaborators in the 1960. See [98]. None of them is really satisfying.

3.5.2 Locality and Separability

Another problem with real algebras comes from the requirement of locality in quantum field theory, and to the related concept of separability of subsystems. Space-time locality will be discussed a bit more later on. But there is already a problem with real algebras when one wants to characterize the properties of a composite system out of those of its subconstituents. As far as I know, this was first pointed out by Araki, and recovered by various people, for instance by Wooter in [111] (see Auletta [8], p. 174, 10.1.3).

Let me consider a system S which consists of two separated subsystem S_1 and S_2. Note that in QFT a subsystem is defined by its subalgebra of observables and of states. These are for instance the "system" generated by the observables in two causally separated regions. Then the algebra of observables $\mathcal{A}$ for the total system $1 + 2$ is the tensor product of the two algebras $\mathcal{A}_1$ and $\mathcal{A}_2$

$$\mathcal{A} = \mathcal{A}_1 \otimes \mathcal{A}_2 \tag{3.36}$$

which means that $\mathcal{A}$ is generated by the linear combinations of the elements $\mathbf{a}$ of the form $\mathbf{a}_1 \otimes \mathbf{a}_2$.

Let me now assume that the algebras of observables $\mathcal{A}_1$ and $\mathcal{A}_2$ are (sub)algebras of the algebra of operators on some *real* Hilbert spaces $\mathcal{H}_1$ and $\mathcal{H}_2$. The Hilbert space of the whole system is the tensor product $\mathcal{H} = \mathcal{H}_1 \otimes \mathcal{H}_2$. Observables are represented by operators A, and physical (symmetric) operators $\mathbf{a} = \mathbf{a}^*$ correspond to symmetric operators $A = A^T$. Now it is easy to see that the physical (symmetric) observables of the whole system are generated by the products of pairs of observables(A_1, A_2) of the two subsystems which are of the form

$$A_1 \otimes A_2 \quad \text{such that} \quad \begin{cases} A_1 \text{ and } A_2 \text{ are both symmetric} \\ \qquad\qquad \text{or} \\ A_1 \text{ and } A_2 \text{ are both skew-symmetric} \end{cases} \tag{3.37}$$

In both case the product is symmetric, but these two cases do not generate the same observables. This is different from the case of algebras of operators on *complex* Hilbert spaces, where all symmetric operators on $\mathcal{H} = \mathcal{H}_1 \otimes \mathcal{H}_2$ are generated by the tensor products of the form

$$A_1 \otimes A_2 \quad \text{such that } A_1 \text{ and } A_2 \text{ are symmetric} \tag{3.38}$$

In other word, if a quantum system is composed of two independent subsystems, and the physics is described by a real Hilbert space, there are physical observables of the big system which cannot be constructed out of the physical observables of the two subsystems! This would turn into a problem with locality, since one could not characterize the full quantum state of a composite system by combining the results of separate independent measurements on its subparts. Note that this is also related to the idea of quantum tomography.

3.5.3   Quaternionic Hilbert Spaces

There has been also serious attempts to build quantum theories (in particular of fields) based on quaternionic Hilbert spaces, both in the '60 and more recently by S. Adler [1]. One idea was that the SU(2) symmetry associated to quaternions could be related to the symmetries of the quark model and of some gauge interaction models. These models are also problematic. In this case there are less physical observables for a composite system that those one can naively construct out of those of the subsystems, in other word there are many non trivial constraints to be satisfied. A far as I know, no satisfying theory based on $\mathbb{H}$, consistent with locality and special relativity, has been constructed.

3.6   Superselection Sectors

3.6.1   Definition

In the general infinite dimensional complex case the decomposition of an algebra of observables $\mathcal{A}$ along its center $Z(\mathcal{A})$ goes in a similar way as in the finite dimensional case. One can write something like

$$\mathcal{A} = \int_{c \in \mathcal{A}'} \mathcal{A}_c \qquad (3.39)$$

where each $\mathcal{A}_c$ is a simple C*-algebra.

A very important difference with the finite dimensional case is that an infinite dimensional C*-algebra $\mathcal{A}$ has in general many inequivalent irreducible representations in a Hilbert space. Two different irreducible representations π_1 and π_2 of $\mathcal{A}$ in two subspaces $\mathcal{H}_1$ and $\mathcal{H}_2$ of a Hilbert space $\mathcal{H}$ are generated by two unitarily inequivalent pure states φ_1 and φ_2 of $\mathcal{A}$. Each irreducible representation π_i and the associated Hilbert space $\mathcal{H}_i$ is called a *superselection sector*. The great Hilbert space $\mathcal{H}$ generated by all the unitarily inequivalent pure states on $\mathcal{A}$ is the direct sum of all superselections sectors. The operators in $\mathcal{A}$ do not mix the different superselection sectors. It is however often very important to consider the operators in $\mathcal{B}(\mathcal{H})$ which mixes the different superselection sectors of $\mathcal{A}$ while respecting the structure of the algebra $\mathcal{A}$ (i.e. its symmetries). Such operators are called *intertwinners*.

3.6.2   A Simple Example: The Particle on a Circle

One of the simplest examples of superselection sector is the nonrelativistic particle on a one dimensional circle. Let us first consider the particle on a line. The two conjugate operators **Q** and **P** obey the canonical commutation relations

$$[\mathbf{Q}, \mathbf{P}] = i \tag{3.40}$$

They are unbounded, but their exponentials

$$\mathbf{U}(k) = \exp(ik\mathbf{Q}) \quad , \qquad \mathbf{V}(x) = \exp(ix\mathbf{Q}) \tag{3.41}$$

generates a C*-algebra. Now a famous theorem by Stone and von Neumann states that all representations of their commutation relations are unitary equivalent. In other word, there is only one way to quantize the particle on the line, given by canonical quantization and the standard representation of the operators acting on the Hilbert space of functions on $\mathbb{R}$.

$$\mathbf{Q} = x \quad , \qquad \mathbf{P} = \frac{1}{i}\frac{\partial}{\partial x} \tag{3.42}$$

Now, if the particle is on a circle with radius 1, the position x becomes an angle θ defined mod. 2π. The operator $\mathbf{U}(k)$ is defined only for integer momenta $k = 2\pi n$, $n \in \mathbb{Z}$. The corresponding algebra of operators has now inequivalent irreducible representations, indexed by a number Φ. Each representation π_Φ corresponds to the representation of the **Q** and **P** operators acting on the Hilbert space $\mathcal{H}$ of functions $\psi(\theta)$ on the circle as

$$\mathbf{Q} = \theta \quad , \qquad \mathbf{P} = \frac{1}{i}\frac{\partial}{\partial \theta} + A \quad , \qquad A = \frac{\Phi}{2\pi} \tag{3.43}$$

So each superselection sector describes the quantum dynamics of a particle with unit charge $e = 1$ on a circle with a magnetic flux Φ. No global unitary transformation (acting on the Hilbert space of periodic functions on the circle) can map one superselection sector onto another one. Indeed this would correspond to the unitary transformation

$$\psi(\theta) \rightarrow \psi(\theta)\, e^{i\theta\,\Delta A} \tag{3.44}$$

and there is a topological obstruction if ΔA is not an integer. Here the different superselection sectors describe different "topological phases" of the same quantum system.

This is of course nothing but the famous Aharonov-Bohm effect [2,42]. Note that this formulation of the Aharonov-Bohm effect in the algebraic formalism does not contradict the usual formulation in the standard formulation in term of a change of boundary conditions for the wave-functions in real space.

3.6.3 General Discussion

The notion of superselection sector was first introduced by Wick, Wightman and Wigner in 1952. They observed (and proved) that is meaningless in a quantum field theory like QED to speak of the superposition of two states ψ_1 and ψ_2 with integer and half integer total spin respectively, since a rotation by 2π changes by (-1) the relative phase between these two states, but does not change anything physically. This apparent paradox disappear when one realizes that this is a similar situation than above. No physical observable allows to distinguish a linear superposition of two states in different superselection sectors, such as $|1 \text{ fermion}\rangle + |1 \text{ boson}\rangle$ from a statistical mixture of these two states $|1 \text{ fermion}\rangle\langle 1 \text{ fermion}|$ and $|1 \text{ boson}\rangle\langle 1 \text{ boson}|$. Indeed, any operator creating or destroying just one fermion is not a physical operator (but rather what is called an intertwining operator). Of course any operator creating or destroying a pair of fermions (or better a pair fermion-antifermion) is a physical operator.

The use of superselection sectors have been sometimes criticized in high energy physics (see a discussion on its use in relation with continuous symmetries in [110]). Superselection sectors are nevertheless a very important feature of the mathematical formulation of quantum field theories, but they do have also physical significance. One encounters superselection sectors in quantum systems with an infinite number of states (non-relativistic or relativistic) as soon as

- the system may be in different phases (for instance in a statistical quantum system with spontaneous symmetry breaking);
- the system has global or local gauge symmetries and sectors with different charges Q_a (abelian or non abelian);
- the system contains fermions;
- the system exhibits different inequivalent topological sectors, this includes the simple case of a particle on a ring discussed above (the Aharonov-Bohm effect), but also gauge theories with θ-vacua;
- more generally, at a mathematical level, a given QFT for different values of its couplings or the masses of particle may corresponds to different superselection sectors of the same operator algebra (Haag's theorem).
- superselection sectors have also been used to discuss measurements in quantum mechanics and the quantum-to-classical transition.

Thus one should keep in mind that the abstract algebraic formalism contains as a whole the different possible states, phases and dynamics of a quantum system, while a given representation describes a subclass of states or of possible dynamics.

3.7   von Neumann Algebras

A special class of C^*-algebras, the so-called von Neumann algebras or W^*-algebras, is of special interest in mathematics and for physical applications. As far as I know these were the algebras of operators originally studied by Murray and von Neumann (the ring of operators). Here I just give some definitions and some motivations, without details or applications.

3.7.1 Definitions

There are several equivalent definitions, I give here three classical definitions. The first two refer to an explicit representation of the algebra as an algebra of operators on a Hilbert space, but the definition turns out to be independent of the representation. The third one depends only on the abstract definition of the algebra.

Weak Closure $\mathcal{A}$ a unital *-sub algebra of the algebra of bounded operators $\mathcal{L}(\mathcal{H})$ on a complex Hilbert space $\mathcal{H}$ is a W*-algebra iff $\mathcal{A}$ is closed under the weak topology, namely if for any sequence A_n in $\mathcal{A}$, when all the individual matrix elements $\langle x|A_n|y\rangle$ converge towards some matrix element A_{xy}, this defines an operator in the algebra

$$\forall \, x, y \, \in \mathcal{H} \quad \langle x|A_n|y\rangle \to A_{xy} \quad \Longrightarrow \quad A \in \mathcal{A} \quad \text{such that} \quad \langle x|A|y\rangle = A_{xy} \tag{3.45}$$

NB: The weak topology considered here can be replaced in the definition by stronger topologies on $\mathcal{L}(\mathcal{H})$. In the particular case of commutative algebras, one can show that W*-algebras correspond to the set of measurable functions $L^\infty(X)$ on some measurable space X, while C*-algebras corresponds to the set $C_0(Y)$ of continuous functions on some Hausdorff space Y. Thus, as advocated by A. Connes, W*-algebras corresponds to non-commutative measure theory, while C*-algebras to non-commutative topology theory.

The Bicommutant Theorem A famous theorem by von Neumann states that $\mathcal{A} \subset L(\mathcal{H})$ is a W*-algebra iff it is a C*-algebra and it is equal to its bicommutant

$$\mathcal{A} = \mathcal{A}'' \tag{3.46}$$

(the commutant $\mathcal{A}'$ of $\mathcal{A}$ is the set of operators that commute with all the elements of $\mathcal{A}$, and the bicommutant is the commutant of the commutant).

NB: The equivalence of this "algebraic" definition with the previous "topological" or "analytical" one illustrate the deep relation between algebra and analysis at work in operator algebras and in quantum physics. It is often stated that this property means that a W*-algebra $\mathcal{A}$ is a symmetry algebra (since $\mathcal{A}$ is the algebras of symmetries of $\mathcal{B} = \mathcal{A}'$). But one can also view this as the fact that a W*-algebra is a "causally complete" algebra of observables, in analogy with the notion of causally complete domain (see the next section on algebraic quantum field theory).

The Predual Property It was shown by Sakai that W*-algebras can also be defined as C*-algebras that have a predual, i.e. when considered as a Banach vector space, $\mathcal{A}$ is the dual of another Banach vector space $\mathcal{B}$ ($\mathcal{A} = \mathcal{B}^\star$).

NB: This definition is unique up to isomorphisms, since $\mathcal{B}$ can be viewed as the set of all (ultra weak) continuous linear functionals on $\mathcal{A}$, which is generated by the positive normal linear functionals on $\mathcal{A}$ (i.e. the states) with adequate topology. So W*-algebras are also algebras with special properties for their states.

3.7.2 Classification of Factors

Let me say a few words on the famous classification of factors. Factors are W*-algebras with trivial center $C = \mathbb{C}$ and any W*-algebra can be written as an integral sum over factors. W*-algebra have the property that they are entirely determined by their projectors elements (a projector is such that $\mathbf{a} = \mathbf{a}^* = \mathbf{a}^2$, and corresponds to orthogonal projections onto closed subspaces E of $\mathcal{H}$). The famous classification result of Murray and von Neumann states that there are basically three different classes of factors, depending on the properties of the projectors and on the existence of a trace.

Type I A factor is of type I if there is a minimal projector E such that there is no other projector F with $0 < F < E$. Type I factors always corresponds to the whole algebra of bounded operators $L(\mathcal{H})$ on some (separable) Hilbert space $\mathcal{H}$. Minimal projector are projectors on pure states (vectors in $\mathcal{H}$). This is the case usually considered by "ordinary physicists". They are denoted I_n if $\dim(\mathcal{H}) = n$ (matrix algebra) and I_∞ if $\dim(\mathcal{H}) = \infty$.

Type II Type II factors have no minimal projectors, but finite projectors, i.e. any projector E can be decomposed into $E = F + G$ where E, F and G are equivalent projectors. The type II_1 hyperfinite factor has a unique finite trace ω (a state such that $\omega(\mathbf{1}) = 1$ and $\omega(\mathbf{aa}^*) = \omega(\mathbf{a}^*\mathbf{a})$), while type $II_\infty = II_1 \otimes I_\infty$. They play an important role in non-relativistic statistical mechanics of infinite systems, the mathematics of integrable systems and CFT.

Type III This is the most general class. Type III factors have no minimal projectors and no trace. They are more complicated. Their classification was achieved by A. Connes. These are the general algebras one must consider in relativistic quantum field theories.

3.7.3 The Tomita-Takesaki Theory

Let me say a few words on an important feature of von Neumann algebras, which states that there is a natural "dynamical flow" on these algebras induced by the states. This will be very sketchy and naive. We have seen that in "standard quantum mechanics" (corresponding to a type I factor), the evolution operator $U(t) = \exp(-itH)$ is well defined in the lower half plane $\mathrm{Im}(t) \leq 0$.

This correspondence "state $\leftrightarrow$ dynamics" can be generalized to any von Neumann algebra, even when the concept of density matrix and trace is not valid any more. Tomita and Takesaki showed that to any state ϕ on $\mathcal{A}$ (through the GNS construction $\phi(\mathbf{a}) = \langle \Omega | \mathbf{a}\Omega \rangle$ where Ω is a separating cyclic vector of the Hilbert space $\mathcal{H}$) one can associate a one parameter family of modular automorphisms σ_t^Φ: $\mathcal{A} \to \mathcal{A}$, such that $\sigma_t^\Phi(\mathbf{a}) = \Delta^{it}\mathbf{a}\Delta^{-it}$, where Δ is positive selfadjoint modular operator in $\mathcal{A}$. This group depends on the choice of the state ϕ only up to inner automorphisms, i.e. unitary transformations u_t such that $\sigma_t^\Psi(\mathbf{a}) = u_t \sigma_t^\Phi(\mathbf{a}) u_t^{-1}$, with the 1-cocycle property $u_{s+t} = u_s \sigma_s(u_t)$.

As advocated by A. Connes, this means that there is a "global dynamical flow" acting on the von Neumann algebra $\mathcal{A}$ (modulo unitaries reflecting the choice of initial state). This Tomita-Takesaki theory is a very important tool in the mathematical theory of operator algebras. It has been speculated by some authors that there is a deep connection between statistics and time (the so called "thermal time hypothesis"), with consequences in quantum gravity. Without defending or discussing more this hypothesis, let me just state that the theory comforts the point of view that operator algebras have a strong link with causality.

3.8 Locality and Algebraic Quantum Field Theory

Up to now I have not really discussed the concepts of time and of dynamics, and the role of relativistic invariance and locality in the quantum formalism. One should remember that the concepts of causality and of reversibility are already incorporated within the formalism from the start.

It is not really meaningful to discuss these issues if not in a fully relativistic framework. This is the object of algebraic and axiomatic quantum field theory. Since I am not a specialist I give only a very crude and very succinct account of this formalism and refer to the excellent book by R. Haag [55] for all the details and the mathematical concepts.

3.8.1 Algebraic Quantum Field Theory in a Dash

In order to make the quantum formalism compatible with special relativity, one needs three things.

Locality Firstly the observables must be built out of the local observables, i.e. the observables attached to bounded domains $\mathcal{O}$ of Minkovski space-time $M = \mathbb{R}^{1,d-1}$. They corresponds to measurements made by actions on the system in a finite region of space, during a finite interval of time. Therefore one associate to each domain $\mathcal{O} \subset M$ a subalgebra $\mathcal{A}(\mathcal{O})$ of the algebra of observables.

$$\mathcal{O} \to \mathcal{A}(\mathcal{O}) \subset \mathcal{A} \tag{3.47}$$

This algebra is such that is

$$\mathcal{A}(\mathcal{O}_1 \cup \mathcal{O}_2) = \mathcal{A}(\mathcal{O}_1) \vee \mathcal{A}(\mathcal{O}_2) \tag{3.48}$$

where $\vee$ means the union of the two subalgebras (the intersection of all subalgebras containing both $A(\mathcal{O}_1)$ and $A(\mathcal{O}_2)$) (Fig. 3.1).

Note that this implies

$$\mathcal{O}_1 \subset \mathcal{O}_2 \implies A(\mathcal{O}_1) \subset A(\mathcal{O}_2) \tag{3.49}$$

The local operators are obtained by taking the limit when a domain reduces to a point (this is not a precise or rigorous definition, in particular in view of the UV divergences of QFT and the renormalization problems).

Caution, the observables of two disjoint domains are not independent if these domains are not causally independent (see below) since they can be related by dynamical/causal evolution (Fig. 3.2).

Causality Secondly causality and locality must be respected, this implies that physical local observables which are causally independent must always commute. Indeed the result of measurements of causally independent observables is always independent of the order in which they are performed, independently of the state of the system. Were this not the case, the observables would not be independent and through some measurement process information could be manipulated and transported at a faster than light pace. If $\mathcal{O}_1$ and $\mathcal{O}_2$ are causally separated (i.e. any $x_1 - x_2$, $x_1 \in \mathcal{O}_1$, $x_2 \in \mathcal{O}_2$ is space-like)) then any pair of operators A_1 and A_2 respectively in $A(\mathcal{O}_1)$ and $A(\mathcal{O}_2)$ commutes

$$\mathcal{O}_1 \bigtimes \mathcal{O}_2, \quad A_1 \in A(\mathcal{O}_1), \quad A_2 \in A(\mathcal{O}_2) \quad \implies \quad [A_1, A_2] = 0 \tag{3.50}$$

This is the crucial requirement to enforce locality in the quantum theory.

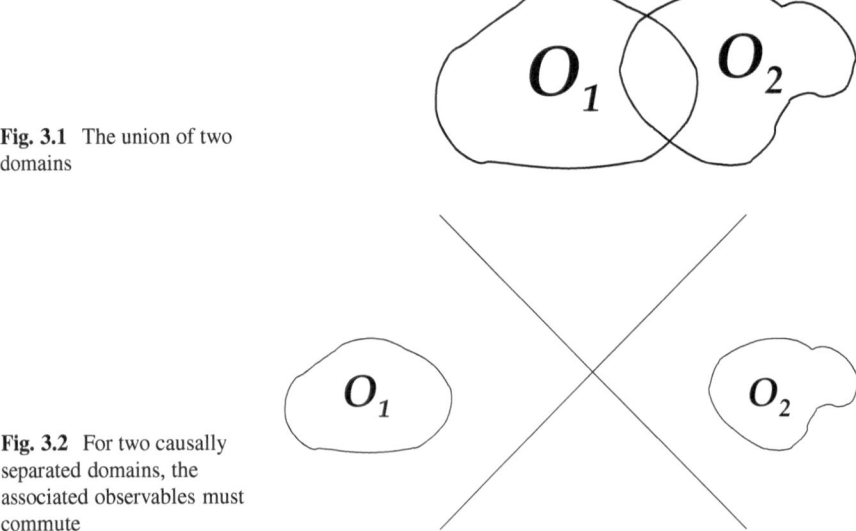

Fig. 3.1 The union of two domains

Fig. 3.2 For two causally separated domains, the associated observables must commute

This indicates also why one should concentrate on von Neumann algebras. The set of local subalgebras $\mathcal{L} = \{\mathcal{A}(\mathcal{O}) : \mathcal{O}$ subdomains of $M\}$ form an orthocomplemented lattice of algebras (see the next chapter to see what this notion is) with interesting properties.

Poincaré Invariance The Poincaré group $\mathfrak{P}(1, d - 1) = \mathbb{R}^{1,d-1} \rtimes O(1, d - 1)$ must act on the space of local observables, so that it corresponds to a symmetry of the theory (the theory must be covariant under translations in space and time and Lorentz transformations). When $\mathcal{A}$ is represented as an algebra of operators on a Hilbert space, the action is usually represented by unitary[4] transformations $U(a, \Lambda)$ (a being a translation and Λ a Lorentz transformation). This implies in particular that the algebra associated to the image of a domain by a Poincaré transformation is the image of the algebra under the action of the Poincaré transformation (Fig. 3.5).

$$U(a, \Lambda)\mathcal{A}(\mathcal{O})U^{-1}(a, \Lambda) = \mathcal{A}(\Lambda\mathcal{O} + a) \tag{3.54}$$

The generator of time translations will be the Hamiltonian $P_0 = H$, and time translations acting on observables corresponds to the dynamical evolution of the system in the Heisenberg picture, in a given Lorentzian reference frame.

The Vacuum State Finally one needs to assume the existence (and the uniqueness, in the absence of spontaneous symmetry breaking) of a special state, the vacuum state $|\Omega\rangle$. The vacuum state must be invariant under the action of the Poincaré transformations, i.e. $U(a, \Lambda)|\Omega\rangle = |\Omega\rangle$. At least in the vacuum sector, the spectrum of $\mathbf{P} = (E, \vec{P})$ (the generators of time and space translations) must lie in the future cone.

$$E^2 - \vec{p}^2 > 0 \quad , \qquad E > 0 \tag{3.55}$$

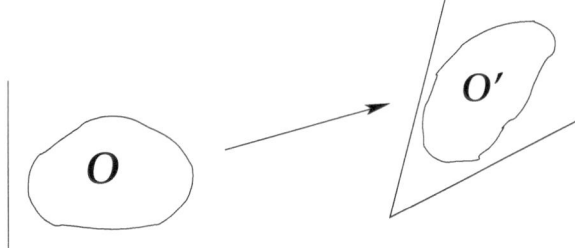

Fig. 3.5 The Poincaré group acts on the domains and on the associated algebras

---

[4]Unitary with respect to the real algebra structure, i.e. unitary or antiunitary w.r.t. the complex algebra structure.

NB: As already discussed, in theories with fermion, fermionic field operators like ψ and $\bar{\psi}$ are not physical operators, since they intertwin different sectors (the bosonic and the fermionic one) and hence the anticommutation of fermionic operators does not contradict the above rule.

Causal Completion One needs also to assume causal completion, i.e.

$$A(\mathcal{O}) = A(\hat{\mathcal{O}}) \tag{3.51}$$

where the domain $\hat{\mathcal{O}}$ is the causal completion of the domain $\mathcal{O}$ ($\hat{\mathcal{O}}$ is defined as the set of points $\mathcal{O}''$ which are causally separated from the points of $\mathcal{O}'$, the set of points causally separated from the points of $\mathcal{O}$, see Fig. 3.3 for a self explanating illustration).

   This implies in particular that the whole algebra A is the (inductive) limit of the subalgebras generated by an increasing sequence of bounded domains whose union is the whole Minkovski space

$$\mathcal{O}_i \subset \mathcal{O}_j \text{ if } i < j \text{ and } \bigcup_i \mathcal{O}_i = \mathbb{M}^4 \quad\Longrightarrow\quad \varinjlim A(\mathcal{O}_i) = A \tag{3.52}$$

and also that it is equal to the algebra associated to "time slices" with arbitrary small time width (Fig. 3.4).

$$S_\epsilon = \{\mathbf{x} = (t, \vec{x}) : t_0 < t < t_0 + \epsilon\} \tag{3.53}$$

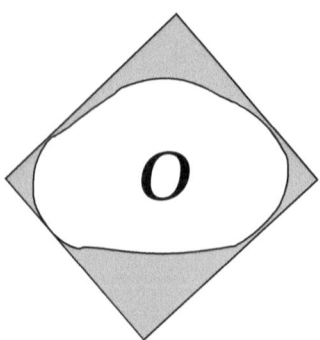

Fig. 3.3 A domain $\mathcal{O}$ and its causal completion $\hat{\mathcal{O}}$ (in *gray*)

Fig. 3.4 An arbitrary thin space-like slice of space-time is enough to generate the algebra of observables A

This is required since the dynamics of the quantum states must respect causality. In particular, the condition $E > 0$ (positivity of the energy) implies that dynamical evolution is compatible with the modular automorphisms on the algebra of observables constructed by the Tomita-Takesaki theory.

3.8.2 Axiomatic QFT

3.8.2.1 Wightman Axioms

One approach to implement the program of algebraic local quantum field theory is the so-called axiomatic field theory framework (Wightman and Gårding). Actually the axiomatic field theory program was started before the algebraic one. In this formalism, besides the axioms of local AQFT, the local operators are realized as "local fields". These local fields Φ are represented as distributions (over space-time M) whose values, when applied to some C^∞ test function with compact support f (typically inside some $\mathcal{O}$) are operators $\mathbf{a} = \langle \Phi \cdot f \rangle$. Local fields are thus "operator valued distributions". They must satisfy the Wightman's axioms (see Streater and Wightman's book [97] and R. Haag's book, again), which enforce causality, locality, Poincaré covariance, existence (and uniqueness) of the vacuum (and eventually in addition asymptotic completeness, i.e. existence of a scattering S-matrix).

3.8.2.2 CPT and Spin-Statistics Theorems

The axiomatic framework is very important for the definition of quantum theories. It is within this formalism that one can derive the general and fundamental properties of relativistic quantum theories

- Reconstruction theorem: reconstruction of the Hilbert space of states from the vacuum expectation values of product of local fields (the Wightman functions, or correlation functions),
- Derivation of the analyticity properties of the correlation functions with respect to space-time $\mathbf{x} = (t, \vec{x})$ and impulsion $\mathbf{p} = (E, \vec{p})$ variables,
- Analyticity of the S matrix (an essential tool),
- The CPT theorem: locality, Lorentz invariance and unitarity imply CPT invariance,
- The spin statistics theorem,
- Definition of quantum field theories in Euclidean time (Osterwalder-Schrader axioms) and rigorous formulation of the mapping between Euclidean theories and Lorentzian quantum theories.

3.9   Discussion

I gave here a short introduction to the algebraic formulation of quantum mechanics and quantum field theory. I did not aim at mathematical rigor nor completeness. I have not mentioned recent developments and applications in the direction of gauge theories, of two dimensional conformal field theories, of quantum field theory in non trivial (but classical) gravitational background.

However I hope to have conveyed the idea that the "canonical structure of quantum mechanics"—complex Hilbert space of states, algebra of operators, Born rule for probabilities—is quite natural and is a representation of an underlying more abstract structure: a real algebra of observables + states, consistent with the physical concepts of causality, reversibility and locality/separability.

Chapter 4
The Quantum Logic Formalism

4.1 Introduction

4.1.1 Why an Algebraic Structure?

The quantum logic formalism is another interesting, albeit more abstract, way to formulate quantum physics. The bonus of this approach is that one does not need to start from the assumption that the set of observables of a physical system is embodied with the algebraic structure of an associative unital algebra. As discussed in the previous section, this assumption that one can "add" and "multiply" observables is already a highly non trivial one. This algebraic structure is natural in classical physics since the observables form a commutative Poisson algebra, addition and multiplication of observables reflect the action of adding and multiplying results of different measurements (it is the Poisson bracket structure that is non trivial). In quantum physics these addition and multiplication operations on observables versus results are not equivalent anymore. Measuring the observable $C = A + B$ does not amount to measure independently the observables A and B and simply add the results, since in general the operators A and B do not commute, and cannot be measured independently. However we have seen for instance that the GNS construction relates the algebra structure of the observables to the Hilbert space structure of the pure states. In particular the superposition principle for pure states is a consequence of the existence of an addition law for the physical observables.

As will be explained in this chapter, in the quantum logic formulations the algebraic structure of the observables (the fact that they form an associative but non-commutative algebra) comes out somehow naturally from more basic assumptions one makes on the measurement operations, or tests, that one can make on physical systems. In the algebraic formulation the Hilbert space of states is reconstructed from the algebra structure of observables. Likewise in the quantum logic formulation this algebraic structure for the observables and the Hilbert space

© Springer International Publishing Switzerland 2015
F. David, *The Formalisms of Quantum Mechanics*, Lecture Notes in Physics 893,
DOI 10.1007/978-3-319-10539-0_4

structure for the states can be derived (with some assumptions) from the possible symmetries that the measurement operations must satisfy.

4.1.2   Measurements as "Logical Propositions"

The idea at the root of the quantum logic approach goes back to J. von Neumann's book [105, 106] and the article by G. Birkhoff[1] and J. von Neumann (again!) [16]. It starts from the remark that the observables given by projectors, i.e. operators $\mathbf{P}$ such that $\mathbf{P}^2 = \mathbf{P} = \mathbf{P}^\dagger$, correspond to propositions with YES or NO (i.e. TRUE or FALSE) outcome in a logical system. An orthogonal projector $\mathbf{P}$ onto a linear subspace $P \subset \mathcal{H}$ is indeed the operator associated to an observable that can take only the values 1 (and always 1 if the state $\psi \in P$ is in the subspace P) or 0 (and always 0 if the state $\psi \in P^\perp$ belongs to the orthogonal subspace to P). Thus we can consider that measuring the observable $\mathbf{P}$ is equivalent to perform a test on the system, or to check the validity of a logical proposition $\mathbf{p}$ on the system.

$$\mathbf{P} = \text{orthogonal projector onto } P \quad \leftrightarrow \quad \text{proposition } \mathbf{p} \qquad (4.1)$$

If the result is 1 the proposition $\mathbf{p}$ is found to be TRUE, and if the result is 0 the proposition $\mathbf{p}$ is found to be FALSE.

$$\langle \psi | \mathbf{P} | \psi \rangle = 1 \implies \mathbf{p} \text{ always TRUE on } | \psi \rangle \qquad (4.2)$$

The projector $\mathbf{1} - \mathbf{P}$ onto the orthogonal subspace $P^\perp$ is associated to the proposition **not p**, meaning usually that $\mathbf{p}$ is false (assuming the law of excluded middle)

$$\langle \psi | \mathbf{P} | \psi \rangle = 0 \implies \mathbf{p} \text{ always FALSE on } | \psi \rangle \qquad (4.3)$$

so that

$$\mathbf{1} - \mathbf{P} = \text{orthogonal projector onto } P^\perp \quad \leftrightarrow \quad \text{proposition not } \mathbf{p} \qquad (4.4)$$

In classical logic the negation **not** is denoted in various ways

$$\text{"not a"} = \neg\mathbf{a}, \ \mathbf{a}', \ \bar{\mathbf{a}}, \ \tilde{\mathbf{a}}, \ \sim\mathbf{a} \qquad (4.5)$$

I shall use the first two notations $\neg\mathbf{a}$, $\mathbf{a}'$.

Now if two projectors $\mathbf{A}$ and $\mathbf{B}$ (on two subspaces A and B) commute, they correspond to classically compatible observables A and B (which can be measured independently), and to a pair of propositions $\mathbf{a}$ and $\mathbf{b}$ of standard logic. The projector

---

[1] An eminent mathematician, not to be confused with his father, the famous G.D. Birkhoff of the ergodic theorem.

$\mathbf{C} = \mathbf{AB} = \mathbf{BA}$ on the intersection of the two subspaces $C = A \cap B$ corresponds to the proposition $\mathbf{c} =$ "$\mathbf{a}$ and $\mathbf{b}$" $= \mathbf{a} \wedge \mathbf{b}$. Similarly the projector $\mathbf{D}$ on the linear sum of the two subspaces $D = A + B$ corresponds to the proposition $\mathbf{d} =$ "$\mathbf{a}$ or $\mathbf{b}$" $= \mathbf{a} \vee \mathbf{b}$.

$$A \cap B \quad \leftrightarrow \quad \mathbf{a} \wedge \mathbf{b} = \mathbf{a} \text{ and } \mathbf{b} \quad , \quad A + B \quad \leftrightarrow \quad \mathbf{a} \vee \mathbf{b} = \mathbf{a} \text{ or } \mathbf{b} \quad (4.6)$$

Finally the fact that for subspaces $A \subset B$, i.e. for projectors $\mathbf{AB} = \mathbf{BA} = \mathbf{A}$, is equivalent to state that $\mathbf{a}$ implies $\mathbf{b}$

$$A \subset B \quad \leftrightarrow \quad \mathbf{a} \implies \mathbf{b} \quad (4.7)$$

This is easily extended to a general (possibly infinite) set of *commuting* projectors. Such a set generates a commuting algebra of observables $\mathcal{A}$, which corresponds to the algebra of functions on some classical space X. The set of corresponding subspaces, with the operations of linear sum, intersection and orthocomplementation $(+, \cap, \perp)$, is isomorphic to a Boolean algebra of propositions with $(\vee, \wedge, \neg)$, or to the algebra of characteristic functions on subsets of X. Indeed, this is just a reformulation of "ordinary logic"[2] where characteristics functions of measurable sets (in a Borel σ-algebra over some set X) can be viewed as logical propositions. Classically all the observables of some classical system (measurable functions over its phase space Ω) can be constructed out of the classical propositions on the system (the characteristic functions of measurable subsets of Ω).

In quantum mechanics all physical observables can be constructed out of projectors. For general, not necessarily commuting projectors $\mathbf{A}$ and $\mathbf{B}$ on subspaces A and B one still associate propositions $\mathbf{a}$ and $\mathbf{b}$. The negation $\neg \mathbf{a}$, the "and" (or "meet") $\mathbf{a} \wedge \mathbf{b}$ and the "or" (or "join") $\mathbf{a} \vee \mathbf{b}$ are still defined by the geometrical operations $\perp, \cap$ and $+$ on subspaces given by (4.6). The "implies" or $\implies$ is also defined by the $\subset$ as in (4.7) (Fig. 4.1).

However the fact that in a Hilbert space projectors do not necessarily commute implies that the standard distributivity law of propositions

$$A \wedge (B \vee C) = (A \wedge B) \vee (A \wedge C) \quad \vee = \text{or} \quad \wedge = \text{and} \quad (4.8)$$

does not hold. It is replaced by the weaker condition (A, B, C are the linear subspaces associated to the projectors $\mathbf{A}, \mathbf{B}, \mathbf{C}$)

$$A \cap (B + C) \supset ((A \cap B) + (A \cap C)) \quad (4.9)$$

which corresponds in terms of propositions (projectors) to

$$(\mathbf{a} \wedge \mathbf{b}) \vee (\mathbf{a} \wedge \mathbf{c}) \implies \mathbf{a} \wedge (\mathbf{b} \vee \mathbf{c}) \quad (4.10)$$

---

[2] In a very loose sense, I am not discussing mathematical logic theory.

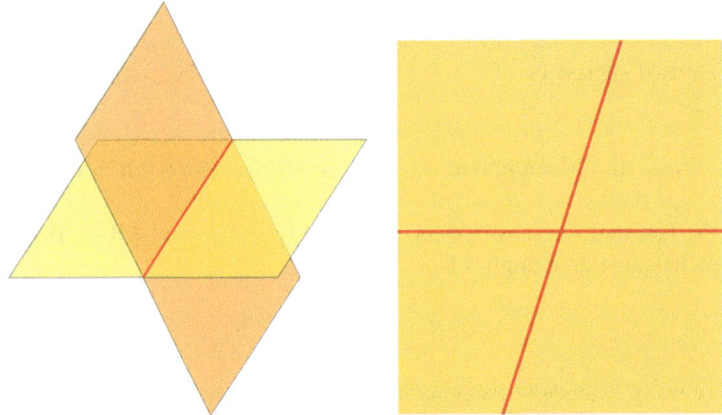

Fig. 4.1 The ∧ (and) as intersection of subspaces, and the ∨ (or) as linear sum of subspaces in quantum logic

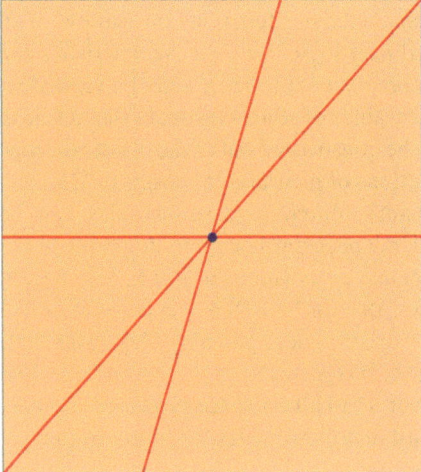

Fig. 4.2 A simple example of non-distributivity

or equivalently

$$\mathbf{a} \vee (\mathbf{b} \wedge \mathbf{c}) \implies (\mathbf{a} \vee \mathbf{b}) \wedge (\mathbf{a} \vee \mathbf{c}) \tag{4.11}$$

A simple example is depicted on Fig. 4.2. The vector space V is the plane (of dimension 2) and the subspaces A, B and C are three different coplanar lines (of dimension 1) passing through the origin. $B + C = V$, hence $A \cap (B + C) = A \cap V = A$, while $A \cap B = A \cap C = \{0\}$; hence $A \cap B + A \cap C = \{0\}$.

Therefore the set of projectors on a Hilbert space do not generate a Boolean algebra, but a more complicated structure, called an orthomodular lattice.

4.1.3 The Quantum Logic Approach

The purpose of the quantum logic approach is to try to understand what are the minimal set of consistency requirements on the propositions/measurements operations, based on logical consistency (assuming that internal consistency has something to do with the physical world) and on physical requirements (in particular the assumptions of causality, reversibility and locality) and what are the consequences of these assumptions for the formulation of physical laws.

This approach was initiated by G. Birkhoff and J. von Neumann (again!) in [16]. It was (slowly) developed by physicists like G. Mackey [77], J.M. Jauch [64] and notably by C. Piron [87, 88], and by mathematicians like Varadarajan[103]. A good reference on the subject (not very recent but very valuable) is the book by E. Beltrametti and G. Cassinelli [15]. It is my main reference and source of understanding

The terminology "quantum logic" for this approach is historical and is perhaps not fully adequate, since it does not mean for most authors that a new kind of logic is necessary to understand quantum physics. It is in fact not a "logic" in the mathematical sense, and it relies on the standard logics used in mathematics and exact sciences. It should rather be called "quantum propositional calculus" or "quantum propositional geometry", where the term "proposition" is to be understood as "test" or "projective measurement" on a quantum system. The mathematics of orthocomplemented and orthomodular lattices that underly the quantum logic formalism have applications in various areas of mathematics, logic and computer sciences.

At that stage I should emphasize that the quantum logic formulation is of no real use for real physical applications! Its formalism is quite heavy, and reduces for practical purposes to the standard formalism of Hilbert spaces and operator algebras, that is better suited and efficient for most if not all uses of the quantum formalism. Its interest is rather when discussing the conceptual foundations.

The quantum logic approaches do not form a unified precise and consistent framework like algebraic quantum field theory. It has several variants, most of them insist on the concept and the properties of propositions (tests), but some older one relying more on the concept of states (in particular the so called convex set approaches). Some more recent formulations of quantum physics related somehow to quantum logic and to quantum information science have some grandiose categorial formulations.

In this course I shall give a short and partial presentation of this approach, from a personal point of view.[3] I shall try to show where the physical concepts of causality, reversibility and locality play a role, in parallel to what I tried to do for the algebraic formalism. My presentation is a classical one based primarily on the concept of propositions. It is also conservative since I not try to use a non-classical logic system (whatever it means) but simply discuss in a classical logic framework the statements which can be made on quantum systems and the measurement operations.

4.2 A Presentation of the Principles

4.2.1 Projective Measurements as Propositions

As explained above, in the standard formulation of quantum mechanics, projectors are associated to "ideal" projective measurements (projective measurement "of the first kind", or "non-demolition" projective measurements). The fundamental property of such measurements is that if the system is already in an eigenstate of the projector, for instance $\mathbf{P}|\psi\rangle = |\psi\rangle$, then after measurement the state of the system is unchanged. This means that successive measurements of P give always the same result (1 or TRUE). Without going into a discussion of measurements in quantum physics, let me stress that this is of course an idealisation of actual measurements. In general physical measurements are not ideal measurements, they may change in a non-minimal way the state of the system. While one gains some information on the system, one in general looses some other information, and measurements may and in general do destroy part or the whole of the system studied. Such general measurements and manipulation processes of a quantum system may be described by the general formalism of POVM's (Projective Operator Valued Measures).

In the following presentation, I assume that such ideal repeatable measurements are in principle possible for all the observable properties of a quantum system. The formalism discussed here will try to understand what is a natural and minimal set of physically reasonable and logically consistent axioms for such measurements.

4.2.2 Causality, POSET's and the Lattice of Propositions

Let me start from a set $\mathcal{L}$ of propositions or tests (associated to ideal measurements) on a physical system, and from a set $\mathcal{E}$ of states (in a similar sense as in the algebraic formulation, to be made more precise along the discussion). On a given state $\varphi \in \mathcal{E}$ the test (measurement) of the proposition a can give TRUE (i.e. YES or 1)

---

[3]with the usual reservation on the lecturer's qualifications.

or FALSE (NO or 0). It gives TRUE with some probability. In this case one has extracted information on the system, which is now (considered to be) in a state φ_a.

I note $\varphi(a)$ the probability that a is found TRUE, assuming that the system was in state φ before the test. I shall not discuss again at that stage what I mean exactly by probability, and refer to the previous discussions.

4.2.2.1 Causal Order Relation

The first ingredient is to assume that there an order relation $a \preceq b$ between propositions. Here it will be defined by the causal relation

$$a \preceq b \quad \Longleftrightarrow \quad \text{for any state } \phi, \text{ if } a \text{ is found true, then } b \text{ will be found true} \tag{4.12}$$

Note that this definition is causal (or dynamical) from the start, as to be expected in quantum physics. It is equivalent to

$$a \preceq b \quad \Longleftrightarrow \quad \forall \phi, \quad \phi_a(b) = 1 \tag{4.13}$$

One assumes that this causal relation has the usual properties of a partial order relation. This amounts to enforce relations between states and propositions. First one must have:

$$a \preceq a \tag{4.14}$$

This means that if a has been found true, the system is now in a state such that a will always be found true. Second one assumes also that

$$a \preceq b \text{ and } b \preceq c \implies a \preceq c \tag{4.15}$$

This is true in particular when, if the system is in a state ψ such that b is always true, then after measuring b, the system is still is the same state ψ. In other word, $\psi(b) = 1 \implies \psi_b = \psi$. This is the concept of repeatability discussed above.

These two properties makes $\preceq$ a *preorder relation* on $\mathcal{L}$.

One also assumes that

$$a \preceq b \text{ and } b \preceq a \implies a = b \tag{4.16}$$

This means that tests which give the same results on any states are indistinguishable. This also means that one can identify a proposition a with the set of states such that a is always found to be true (i.e. $\psi(a) = 1$). Equation (4.16) makes $\preceq$ a partial order relation and $\mathcal{L}$ a *partially ordered set* or POSET.

4.2.2.2   AND (Meet ∧)

The second ingredient is the notion of logical conjunction AND. One assumes that for any pair of test a and b, there is a *unique greater proposition*, denoted $a \wedge b$, such that

$$a \wedge b \preceq a \text{ and } a \wedge b \preceq b \qquad (4.17)$$

in other word, there is a unique $a \wedge b$ such that

$$c \preceq a \text{ and } c \preceq b \implies c \preceq a \wedge b$$

NB: This is a non trivial assumption, not a simple consequence of the previous ones. It can be justified using the notion of filters (see Jauch [64]) or using the notion of questions associated to propositions (see Piron [87]). Here to make the discussion simpler I just present it as an assumption. On the other hand it is very difficult to build anything without this assumption. Note that (4.17) implies (4.16).

This definition extends to any set $\mathcal{A}$ of propositions

$$\bigwedge \mathcal{A} = \bigwedge \{a \in \mathcal{A}\} = \text{greatest } c : c \preceq a, \ \forall a \in \mathcal{A} \qquad (4.18)$$

I do not discuss if the set $\mathcal{A}$ is finite or countable.

4.2.2.3   Logical OR (Join ∨)

From this we can infer the existence of a logical OR (by using Birkhoff theorem)

$$a \vee b = \bigwedge \{c : a \preceq c \text{ and } b \preceq c\} \qquad (4.19)$$

which extents to sets of propositions

$$\bigvee \mathcal{A} = \bigwedge \{b : a \preceq b, \ \forall a \in \mathcal{A}\} \qquad (4.20)$$

4.2.2.4   Trivial 1 and Vacuous ∅ Propositions

It is natural to assume that there is a trivial proposition **1** that is always true

$$\text{for any state } \phi, \ \mathbf{1} \text{ is always found to be true, i.e. } \phi(\mathbf{1}) = 1 \qquad (4.21)$$

and another "vacuous" proposition ∅ that is never true

$$\text{for any state } \phi, \ \emptyset \text{ is never found to be true, i.e. } \phi(\emptyset) = 0 \qquad (4.22)$$

Naturally one has

$$1 = \bigvee \mathcal{L} \text{ and } \emptyset = \bigwedge \mathcal{L} \tag{4.23}$$

With these assumptions and definitions the set of propositions $\mathcal{L}$ has now the structure of a **complete lattice**.

4.2.3 Reversibility and Orthocomplementation

4.2.3.1 Negations a' and 'a

I have not yet discussed what should be considered possible when a proposition is found to be false. To do so one must introduce the seemingly simple notion of negation or complement. In classical logic this is rather easy. The subtle point is that for quantum systems, where causality matters, there are two inequivalent ways to introduce the negation. These two definitions becomes equivalent only if one assumes that propositions on quantum systems share a property that I denote "causal reversibility", by analogy with what I did before, and that I am going to explain. If this property is satisfied, one recovers the standard negation of propositions in classical logic, and ultimately this will lead to the notion of orthogonality and of scalar product of standard quantum mechanics. Thus here again, as in the previous section, reversibility appears to be one of the essential feature of the principles of quantum physics.

Negation—Définition 1 To any proposition a one can associate its negation (or complement proposition) a' defined as

for any state ϕ, if a is found to be true, then a' will be found to be false (4.24)

a' can be defined equivalently as

$$a' = \bigvee \{b \text{ such that on any state } \phi, \text{ if } a \text{ is found true, then } b \text{ will be found false}\} \tag{4.25}$$

Negation—Définition 2 It is important at that stage to realize that, because of the causal ordering in the above definition of a', there is an alternate but symmetric definition for the negation, that I denote $'a$, which is given by

for any state ϕ, if $'a$ is found to be true, then a will be found to be false (4.26)

or equivalently

$$'a = \bigvee \{b \,; \text{ such that on any state } \phi \,, \text{ if } b \text{ is found true, then } a \text{ will be found false}\} \tag{4.27}$$

These two definitions are *not equivalent*, and from the axioms that we choose up to now, each of them do not fulfill the properties of the negation in classical propositional logic.[4]

$$\neg(\neg a) = a \quad \text{and} \quad \neg(a \wedge b) = \neg a \vee \neg b$$

These problems come from the fact that the definition for the causal order $a \preceq b$ does not implies that $b' \preceq a'$, as in classical logic. Indeed the definition (4.12) for $a \preceq b$ implies that for every state

$$\text{if } b \text{ is found false, then } a \text{ was found false} \qquad (4.28)$$

while $b' \preceq a'$ would mean

$$\text{if } b \text{ is found false, then } a \text{ will be found false} \qquad (4.29)$$

or equivalently

$$\text{if } a \text{ is found true, then } b \text{ was found true} \qquad (4.30)$$

4.2.3.2   Causal Reversibility and Negation

In order to build a formalism consistent with what we know of quantum physics, we need to enforce the condition that the causal order structure on propositions is in fact independent of the choice of a causal arrow "if $\cdots$, then $\cdots$ will be $\cdots$" versus "if $\cdots$, then $\cdots$ was $\cdots$" . This is nothing but the requirement of causal reversibility that we discussed before, and it is enforced by the following simple but very important condition.

Causal Reversibility   One assumes that the negation a' is such that

$$a \preceq b \quad \Longleftrightarrow \quad b' \preceq a' \qquad (4.31)$$

With this assumption, it is easy to show that the usual properties of negation are satisfied. The two alternate definitions of negation are now equivalent

$$a' = {}'a = \neg a \qquad (4.32)$$

and may be denoted by the standard logical symbol $\neg$. We then have

$$(a')' = a \qquad (4.33)$$

---

[4]The point discussed here is a priori not connected to the classical versus intuitionist logics debate. Remember that we are not discussing a logical system.

and

$$(a \wedge b)' = a' \vee b' \tag{4.34}$$

as well as

$$\emptyset = \mathbf{1}', \quad \emptyset' = \mathbf{1} \tag{4.35}$$

and

$$a \wedge a' = \emptyset, \quad a \vee a' = \mathbf{1} \tag{4.36}$$

A lattice $\mathcal{L}$ embodied with a complement $\neg$ with the properties (4.31)–(4.36) is called an *orthocomplemented complete lattice* (in short OC lattice). For such a lattice, the couple (a, a') describes what is called a perfect measurement.

NB   Note that in Boolean logic, the implication $\rightarrow$ can be defined from the negation $\neg$. Indeed $a \rightarrow b$ means $\neg a \vee b$. Here it is the negation $\neg$ which is defined out of the implication $\preceq$.

4.2.3.3   Orthogonality

With reversibility and complement, the set of propositions begins to have properties similar to the set of projections on linear subspaces of a Hilbert space.[5] The complement a' of a proposition a is similar to the orthogonal subspace $P^{\perp}$ of a subspace P. This analogy can be extended to the general concept of orthogonality.

Orthogonal Propositions

Two proposition a and b are orthogonal, if $b \preceq a'$ (or equivalently $a \preceq b'$).
$$\tag{4.37}$$

This is noted

$$a \perp b \tag{4.38}$$

Compatible Propositions   OC lattices contain also the concept of family of compatible propositions. A subset of an OC lattice $\mathcal{L}$ is a sublattice $\mathcal{L}'$ if it is stable under the operations $\wedge$, $\vee$ and $'$ (hence it is itself an OC lattice). To any subset $\mathcal{S} \subset \mathcal{L}$ on can associate the sublattice $\mathcal{L}_{\mathcal{S}}$ generated by $\mathcal{S}$, defined as the smallest sublattice $\mathcal{L}'$ of $\mathcal{L}$ which contains $\mathcal{S}$.

---

[5]One should be careful for infinite dimensional Hilbert spaces and general operator algebras. Projectors correspond in general to orthogonal projections on closed subspaces.

A lattice is said to be *Boolean* if it satisfy the distributive law of classical logic

$$a \wedge (b \vee c) = (a \wedge b) \vee (a \wedge c)$$

A subset S of an OC lattice $\mathcal{L}$ is said to be a subset of *compatible propositions* if the generated sublattice $\mathcal{L}_S \subset \mathcal{L}$ is Boolean.

Compatible propositions are the analog of commuting projectors, i.e. compatible or commuting observables in standard quantum mechanics. For a set of compatible propositions, one expects that the expectations of the outcomes YES or NO will satisfy the rules of ordinary logic.

Orthogonal Projection   The notion of orthogonal projection onto a subspace can be also formulated in this framework as

$$\text{projection of } a \text{ onto } b = \Phi_b(a) = b \wedge (a \vee b') \tag{4.39}$$

This projection operation is often called the Sasaki projection. Its dual $(\Phi_b(a'))' = b' \vee (a \wedge b)$ is called the Sasaki hook $(b \overset{S}{\to} a)$. It has the property that even if $a \not\leq b$, if for a state ψ the Sasaki hook $(a \overset{S}{\to} b)$ is always true, then for this state ψ, if a is found true then b will always be found true.

4.2.4   Subsystems of Propositions and Orthomodularity

4.2.4.1   What Must Replace Distributivity?

The concept of orthocomplemented lattice of propositions is not sufficient to reconstruct a consistent quantum formalism. There are mathematical reasons and physical reasons.

One reason is that if the distributive law $A \wedge (B \vee C) = (A \wedge B) \vee (A \wedge C)$ is known not to apply, assuming no restricted distributivity condition is not enough and leads to too many possible structures. In particular in general a lattice with an orthocomplementation $\neg$ may be endowed with other inequivalent orthocomplementations! This is problematic for the physical interpretation of the complement as $a \to$ TRUE $\iff a' \to$ FALSE.

Another problem is that in physics one is led to consider conditional states and conditional propositions. In classical physics this would correspond to the restriction to some subset Ω' of the whole phase space Ω of a physical system, or to the projection $\Omega \to \Omega'$. Such projections or restrictions are necessary if there are some constraints on the states of the system, if one has access only to some subset of all the physical observables of the system, or if one is interested only in the study of a subsystem of a larger system. In particular such a separation of the degrees of freedom is very important when discussing locality: one is interested in the properties of the system we can associate to (the observables measured in) a given interval of space and time, as already discussed for algebraic QFT. It is also very important when discussing effective low energy theories: one wants to separate (project out) the (un-observable) high energy degrees of freedom from the (observable) low energy degrees of freedom. And of course this is crucial to discuss open quantum systems, quantum measurement processes, decoherence processes, and the emergence of classical degrees of freedom and classical behaviors in quantum systems.

4.2.4.2 Sublattices and Weak-Modularity

In general a subsystem is defined from the observables (propositions) on the system which satisfy some constraints. One can reduce the discussion to a single constraint a. If $\mathcal{L}$ is an orthocomplemented lattice and a a proposition of $\mathcal{L}$, let me consider the subset $\mathcal{L}_{<a}$ of all propositions which imply a

$$\mathcal{L}_{<a} = \{b \in \mathcal{L} : b \preceq a\} \tag{4.40}$$

One may also consider the subset of propositions $\mathcal{L}_{>a}$ of propositions implied by a

$$\mathcal{L}_{>a} = \{b \in \mathcal{L} : a \preceq b\} = (\mathcal{L}_{<a'})' \tag{4.41}$$

Is this set of propositions $\mathcal{L}_{<a}$ still an orthocomplemented lattice? Let me take as order relation $\preceq$, $\vee$ and $\wedge$ in $\mathcal{L}_{<a}$ the same than in $\mathcal{L}$ and as trivial and empty propositions $\mathbf{1}_{<a} = a$, $\emptyset_{<a} = \emptyset$. Now, given a proposition $b \in \mathcal{L}_{<a}$, one must define what is its complement $b'_{<a}$ in $\mathcal{L}_{<a}$. A natural choice is

$$b'_{<a} = b' \wedge a \tag{4.42}$$

but in general with such a choice $\mathcal{L}_{<a}$ is not an orthocomplemented lattice, since it is easy to find for general AC lattices counterexamples such that one may have $b \vee b'_{<a} \neq a$.

Weak-Modularity In order for $\mathcal{L}_{<a}$ to be an orthocomplemented lattice (for any $a \in \mathcal{L}$), the orthocomplemented lattice $\mathcal{L}$ must satisfy the weak-modularity condition

$$b \preceq a \implies (a \wedge b') \vee b = a \tag{4.43}$$

This condition is also sufficient.

4.2.4.3 Orthomodular Lattices

Orthomodularity An OC lattice which satisfies the weak-modularity condition is said to be an *orthomodular lattice* (or OM lattice).[6] Clearly if $\mathcal{L}$ is OM, for any $a \in \mathcal{L}$, $\mathcal{L}_{<a}$ is also OM, as well as $\mathcal{L}_{>a}$.

Equivalent Definitions Weak-modularity has several equivalent definitions. Here are two interesting ones:

– $a \preceq b \implies a$ and b are compatible.
– the orthocomplementation $a \to a'$ is unique in $\mathcal{L}$.

Irreducibility For such lattices one can also define the concept of irreducibility. We have seen that two elements a and b of $\mathcal{L}$ are compatible (or commute) if they generate a Boolean lattice. The center $\mathcal{C}$ of a lattice $\mathcal{L}$ is the set of $a \in \mathcal{L}$ which commute with all the elements of the lattice $\mathcal{L}$. It is obviously a Boolean lattice. A lattice is irreducible if its center $\mathcal{C}$ is reduced to the trivial lattice $\mathcal{C} = \{\emptyset, \mathbf{1}\}$.

4.2.4.4 Weak-Modularity Versus Modularity

NB The (somewhat awkward) denomination "weak-modularity" is historical. Following Birkhoff and von Neumann the stronger "modularity" condition for lattices was first considered. Modularity is defined as

$$a \preceq b \implies (a \vee c) \wedge b = a \vee (c \wedge b) \tag{4.44}$$

Modularity is equivalent to weak modularity for finite depth lattices (as a particular case the set of projectors on a finite dimensional Hilbert space for a modular lattice). But modularity turned out to be inadequate for infinite depth lattices (corresponding to the general theory of projectors in infinite dimensional Hilbert spaces). The theory of modular lattices has links with some W*-algebras and the theory of "continuous geometries" (see e.g. [107]).

4.2.5 Pure States and AC Properties

Orthomodular (OM) lattices are a good starting point to consider the constraints that we expect for the set of ideal measurements on a physical system, and therefore to study how one can represent its states. In fact one still needs two more assumptions, which seem technical, but which are also very important (and quite natural from the point of view of quantum information theory). They rely on the concept of atoms,

---

[6]In French: treillis orthomodulaire, in German: Orthomodulare Verband.

or minimal proposition, which are the analog for propositions of the concept of minimal projectors on pure states in the algebraic formalism.

4.2.5.1   Atoms

An element a of an OM lattice is said to be an atom if

$$b \preceq a \quad \text{and} \quad b \neq a \quad \Longrightarrow \quad b = \emptyset \tag{4.45}$$

This means that a is a minimal non empty proposition; it is not possible to find another proposition compatible with a which allows to obtain more information on the system than the information obtained if a is found to be TRUE.

Atoms are the analog of projectors on pure states in the standard quantum formalism (pure propositions). Indeed, if the system in some state ψ, before the measurement of a, if a is an atom and is found to be true, the system will be in a pure state ψ_a after the measure.

4.2.5.2   Atomic Lattices

A lattice is said to be atomic if every non trivial proposition $b \neq \emptyset$ in $\mathcal{L}$ is such that there is at least one atom a such that $a \preceq b$ (i.e. any proposition "contains" at least one minimal non empty proposition). For an atomic OM lattice one can show that any proposition b is then the union of its atoms (atomisticity).

4.2.5.3   Covering Property

Finally one needs also the covering property. The formulation useful in the quantum framework is as follows: An atomic lattice has the covering property if for any proposition a and any atom b not in the complement a' of a, then the Sasaki projection of b onto a, $\Phi_a(b) = a \wedge (b \vee a')$, is still an atom.

The original definition of the covering property for atomic lattices according to Birkhoff is: for any $b \in \mathcal{L}$ and any atom $a \in \mathcal{L}$ such that $a \wedge b = \emptyset$, $a \vee b$ covers b, i.e. there is no c between b and $a \vee b$ such that $b \prec c \prec a \vee b$.

This covering property is very important. It means that when reducing a "system" (or rather a set of states) to a "subsystem" by some constraint (projection onto a), one cannot get a non-minimal proposition on the subset out of a minimal one on the larger set. This would mean that one could get more information out of a subsystem than from the greater system. In other word, if a system is in a pure state, performing a perfect measurement can only map it onto another pure state. Perfect measurements cannot decrease the information on the system.

The covering property is in fact also related to the superposition principle. Indeed, it implies that (for irreducible lattices) for any two difference atoms a and b, there must be a third atom c different from a and b such that $c \prec a \vee b$. Thus, in the weakest possible sense (remember we have no addition law yet) c is a superposition of a and b.

An atomistic lattice with the covering property is said to be an AC lattice. As mentioned before these properties can be formulated in term of the properties of the set of states on the lattice rather than in term of the propositions. This is the object of the convex set approach. I shall not discuss this here.

4.3   The Geometry of Orthomodular AC Lattices

I have given one (possible and personal) presentation of the principles of the quantum logic formalism. It took some time since I tried to explain both the mathematical formalism and the underlying physical ideas. I now explain the main mathematical result: the definition of the set of propositions (ideal measurements) on a quantum system as an orthomodular AC lattice can be equivalently represented as the set of orthogonal projections on some "generalized Hilbert space ".

4.3.1   Prelude: The Fundamental Theorem of Projective Geometry

The idea is to extend a classical and beautiful theorem of geometry, the Veblen-Young theorem. Any abstract projective geometry can be realized as the geometry of the affine subspaces of some left-module (the analog of vector space) on a division ring K (a division ring is a non-commutative generalization of a field like $\mathbb{R}$ or $\mathbb{C}$). This result is known as the "coordinatization of projective geometry". Classical references on geometry are the books by E. Artin *Geometric Algebra* [7], R. Baer *Linear algebra and projective geometry* [11]. More precisely, a geometry on a linear space is simply defined by a set X whose elements x are called the points of X, and a set $\mathcal{L}$ of lines ℓ of X (at that stage a line ℓ is simply a subset of X, so $\mathcal{L}$ is a subset of the set of subsets of X).

Theorem   If the geometry $(X, \mathcal{L})$ satisfies the following axioms:

1. Any line contains at least three points,
2. Two points lie in a unique line,
3. A line meeting two sides of triangle, not at a vertex of the triangle, meets the third side also (Veblen's axiom),
4. There are at least four points non coplanar (a plane is defined in the usual way from lines),

then the corresponding geometry is the geometry of the affine subspaces of a left module M on a division ring K (a division ring is a in general non-commutative field).

Discussion   The theorem here is part of the Veblen-Young theorem, that encompasses the case when the 4th axiom is not satisfied. The first two axioms define a line geometry structure such that lines are uniquely defined by the pairs of points, but with some superposition principle. The third axiom is represented on Fig. 4.3. The fourth axiom is necessary to exclude some special case of non-Desarguesian geometries, that are of no relevance for our discussion.

Let me recall that the division ring K (an associative algebra with an addition $+$, a multiplication $\times$ and an inverse $x \to x^{-1}$) is constructed out of the symmetries of the geometry, i.e. of the automorphisms, or applications $X \to X$, $\mathcal{L} \to \mathcal{L}$, etc. which preserve the geometry. Without giving any details, let me illustrate the case of the standard real projective plane (where $K = \mathbb{R}$). The field structure on $\mathbb{R}$ is obtained by identifying $\mathbb{R}$ with a projective line ℓ which contains the three points 0, 1 and ∞. The "coordinate" $x \in \mathbb{R}$ of a point $X \in \ell$ is identified with the cross-ratio $x = (X, 1; 0, \infty)$. On Fig. 4.4 are depicted the geometrical construction of the addition $X + Y$ and of the multiplication $X \times Y$ of two points X and Y on a line ℓ.

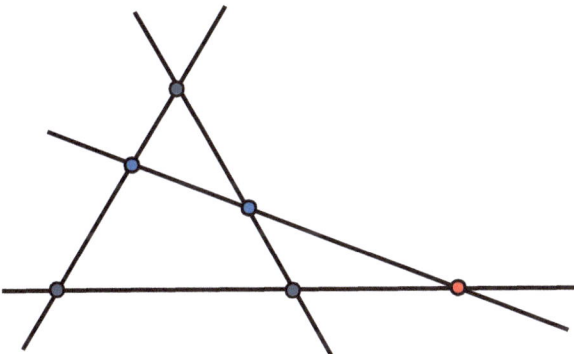

Fig. 4.3 Veblen's axiom

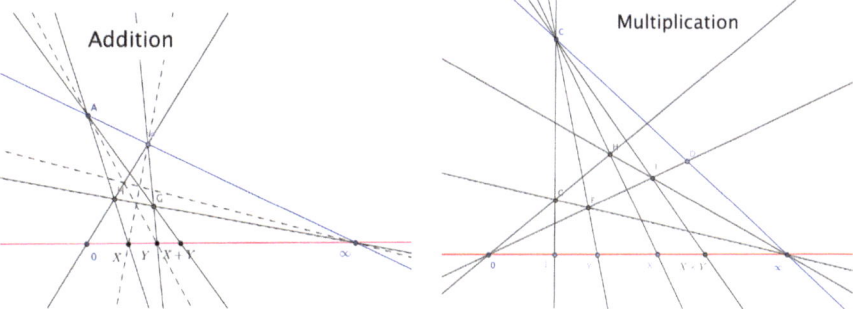

Fig. 4.4 Construction of $+$ and $\times$ in the projective real plane

4.3.2   The Projective Geometry of Orthomodular AC Lattices

4.3.2.1   The Coordinatization Theorem

Similar "coordinatization" theorems hold for the orthomodular AC lattices that
have been introduced in the previous section. The last axioms AC (atomicity and
covering) play a similar role as the axioms of abstract projective geometry, allowing
to define "points" (the atoms), lines, etc , with properties similar to the first 3 axioms
of linear spaces. The difference with projective geometry is the existence of the
orthocomplementation (the negation $\neg$) which allows to define an abstract notion
of orthogonality $\perp$, and the specific property of weak-modularity (which allows to
define in a consistent way what are projections on closed subspaces). Let me first
state the main theorem

Theorem (Piron) Let $\mathcal{L}$ be a complete irreducible orthocomplemented AC lattice
with length > 3 (i.e. at least three 4 different levels of proposition $\emptyset \prec a \prec b \prec c \prec
d \prec 1$). Then the "abstract" lattice $\mathcal{L}$ can be represented as the lattice $\mathcal{L}(V)$ of the
closed subspaces of a left-module[7] V on a division ring[8] K with a Hermitian form
f. The ring K, the module V and the form f have the following properties:

- The division ring K has an involution * such that $(xy)^* = y^* x^*$
- The vector space V has a non degenerate Hermitian (i.e. sesquilinear) form f :
  $V \times V \to K$

$$\mathbf{a}, \mathbf{b} \in V \ , \ \ f(a,b) = \langle \mathbf{a} | \mathbf{b} \rangle \in K \ \ , \ \ \langle \mathbf{a} | \mathbf{b} \rangle = \langle \mathbf{b} | \mathbf{a} \rangle^* \qquad (4.46)$$

- The Hermitian form f defines an orthogonal projection and associates to each
  linear subspace M of V its orthogonal $M^{\perp}$.

$$M^{\perp} = \{ \mathbf{b} \in v : \ \langle \mathbf{b} | \mathbf{a} \rangle = 0 \ \forall \mathbf{a} \in M \} \qquad (4.47)$$

- The closed subspaces of V are the subspaces M such that $(M^{\perp})^{\perp} = M$.
- The Hermitian form is orthomodular, i.e. for any closed subspace, $M^{\perp} + M = V$.
- The OM structure $(\preceq, \wedge, ')$ on the lattice $\mathcal{L}$ is isomorphic to the standard lattice
  structure $(\subseteq, \cap, \perp)$ (subspace of, intersection of, orthogonal complement of)
  over the space $\mathcal{L}(V)$ of closed linear subspaces of V.
- Moreover, V and K are such that there is some element a of V with "norm" unity
  $f(a,a) = 1$ (where 1 is the unit element of K).

I do not give the proof. I refer to the physics literature: [15, Chap. 21], [87, 88],
and to the original mathematical literature [16, 78, 103].

Thus this theorem states that an OM AC lattice can be represented as the lattice
of orthogonal projections over the closed linear subspaces of some "generalized

---

[7] A module is the analog of a vector space, but on a ring instead of a (commutative) field.

[8] A division ring is the analog of a field, but without commutativity.

Hilbert space" with a quadratic form defined over some non-commutative field K. This is very suggestive of the fact that Hilbert spaces are not abstract and complicated mathematical objects, much less natural from a physical point of view than the ordinary numbers and functions of classical mechanics (this statement is still found in some discussion and outreach presentations). On the contrary they are quite natural objects if one wants to describe ideal measurements in quantum physics. In particular the algebra (the underlying ring K and the algebraic structure of the space V) come out naturally from the symmetries of the lattice of propositions $\mathcal{L}$.

4.3.2.2 Discussion: Which Division Ring K?

The important theorem discussed before is very suggestive, but is not sufficient to "derive" standard quantum mechanics. The main question is which division algebra K and which involution * and Hermitian form f are physically allowed? Can one construct physical theories based on other rings than the usual $K = \mathbb{C}$ (or $\mathbb{R}$ or $\mathbb{H}$)?

There is a large variety of division rings ! The simplest one are the finite division rings (with a finite number of elements. The first Wedderburn theorem implies that any finite K is a direct product of Galois fields $\mathbb{F}_p = \mathbb{Z}/\mathbb{Z}_p$ (p prime). Beyond the standard fields $\mathbb{C}$, $\mathbb{R}$ and $\mathbb{H}$, more complicated division rings are the rings of rational functions $F(X)$, and some are even larger (for instance the surreal numbers...), but still commutative,and some are also fully non-commutatives rings.

However, the requirements that K has an involution, and that V has a non degenerate hermitian form, so that $\mathcal{L}(V)$ is a OM lattice, put already very stringent constraints on K. For instance, it is well known that finite fields like the $\mathbb{F}_p$ (p prime) do not work. Indeed, it is easy to see that the lattice $\mathcal{L}(V)$ of the linear subspaces of the finite dimensional subspaces of the n-dimensional vector space $V = (\mathbb{F}_p)^n$ is not orthomodular and cannot be equipped with a non-degenerate quadratic form! Check with $p = n = 3$! But still many more exotic division rings K than the standard $\mathbb{R}$, $\mathbb{C}$ (and $\mathbb{H}$) are possible at that stage.

4.3.3 Towards Hilbert Spaces

There are several arguments that point towards the standard physical solution, V must be a Hilbert space over $\mathbb{R}$, $\mathbb{C}$ (or possibly $\mathbb{H}$). However none of these arguments is fully mathematically convincing, if most would satisfy a physicist. Remember that real numbers are expected to occur in physical theory for two reasons. Firstly the lattice structure is related to probabilities p, which are real numbers. Thus the continuous structure of the real numbers should be contained in the abstract structure (this is not really a very strong argument...). Secondly we look for a quantum theory that must be compatible with the relativistic concept of space-time, where space and time are described by continuous real variables (this is a experimental fact). Of course this is correct as long as one does not try to quantize gravity.

We have not discussed yet precisely the structure of the states ψ, and which constraints they may enforce on the algebraic structure of propositions. Remember that it is the set of states $\mathcal{E}$ which allows to discuss the partial order relation $\preceq$ on the set of propositions $\mathcal{L}$. Moreover states ψ assign probabilities $\psi(a) \in [0, 1]$ to propositions a, with the constraints that if $a \perp b$, $\psi(a \vee b) = \psi(a) + \psi(b)$. Moreover the propositions $a \in \mathcal{L}$ (projective measurements) define via the Sasaki orthogonal projections π_a a set of transformations $\mathcal{L} \to \mathcal{L}$, which form a so called Baer *-semi group. On the same time, propositions $a \in \mathcal{L}$ define mappings $\psi \to \psi_a$ on the states. Since, as in the algebraic formalism, convex linear combinations of states are states, $\mathcal{E}$ generate a linear vector space E, and form a convex subset $\mathcal{E} \subset E$. Thus there is more algebraic structure to discuss than what I explained up to now. I refer to [15], Chap. 16–19, for more details. I shall come back to states when discussing Gleason's theorem in the next section.

Assuming some "natural" continuity or completeness conditions for the states leads to theorems stating that the division ring K must contain the field of real numbers $\mathbb{R}$, hence is $\mathbb{R}$, $\mathbb{C}$ or $\mathbb{H}$, and that the involution * is continuous, hence corresponds to the standard involution $x^* = x$, $x^* = \bar{x}$ or $x^* = x^\star$ respectively. See [15, Chap. 21.3].

Another argument comes from an important theorem in the theory of orthomodular lattices, which holds for lattices of projections in infinite dimensional modules.

Solèr's Theorem (Solèr [95]) Let $\mathcal{L} = \mathcal{L}_K(V)$ be an irreducible OM AC lattice of compact linear suspaces in a left-module V over a division ring K, as discussed above. If there is an infinite family $\{v_i\}$ of orthonormal vectors in V such that $\langle v_i | v_j \rangle = \delta_{ij} f$ with some $f \in K$ then the division ring K can only be $\mathbb{R}$, $\mathbb{C}$ or $\mathbb{H}$.

The proof of this highly non trivial theorem is given in [95]. It is discussed in more details in [61].

The assumptions of the theorem state that there an infinite set of mutually compatible atoms $\{a_i\}_{i \in I}$ in $\mathcal{L}$ (commuting, or causally independent elementary propositions a_i), and in addition that there is some particular symmetry between the generators $v_i \in V$ of the linear spaces (the lines or rays) of these propositions.

The first assumption is quite natural if we take into account space-time and locality in quantum physics. Let me consider the case where the physical space in which the system is defined to be infinite (flat) space or some regular lattice, so that it can be separated into causally independent pieces $\mathcal{O}_\alpha$ (labelled by $\alpha \in \Lambda$ some infinite lattice). See for instant Fig. 4.5. It is sufficient to have one single proposition a_α relative to each $\mathcal{O}_\alpha$ only (for instance "there is one particle in $\mathcal{O}_\alpha$") to build an infinite family of mutually orthogonal propositions $b_\alpha = a_\alpha \wedge (\bigwedge_{\beta \neq \alpha} \neg a_\beta)$ in $\mathcal{L}$. Out of the b_α, thanks to the atomic property (A), we can extract an infinite family of orthogonal atoms c_α.

However this does not ensure the second assumption: the fact that the corresponding $v_i \in V$ are orthonormals. The group of space translations $\mathfrak{T}$ must act as a group of automorphism on the lattice of linear subspaces $\mathcal{L} = \mathcal{L}_K(V)$ (a group

Fig. 4.5 A string of causal diamonds (in space-time)

of automorphisms on a OC lattice $\mathcal{L} = \mathcal{L}_K(V)$ is a group of transformations which preserves the OC lattice structure ($\preceq$, $\wedge$, $'$) or equivalently ($\subseteq$, $\cap$, $\perp$)). There must correspond an action (a representation) of the translation group $\mathfrak{T}$ on the vector space V, and on the underlying field K. If the action is trivial the conditions of Soler's theorem are fulfilled, but this is not ensured a priori. See for instance [46] for a recent discussion of symmetries in orthomodular geometries. However I am not aware of a counterexample where a non standard orthomodular geometry (i.e. different from that of a Hilbert space on $\mathbb{C}$ (or $\mathbb{R}$) carries a representation of a "physical" symmetry group such as the Poincaré or the Galilean group of space-time transformations (representations of these groups should involve the field of real numbers $\mathbb{R}$ in some form).

From now on we assume that a quantum system may indeed be described by projectors in a real or complex Hilbert space.

One last remark. The coordinatization theorem depends crucially on the fact that the OM lattice $\mathcal{L}$ is atomic, hence contain minimal propositions (atoms). They are the analog of minimal projectors in the theory of operator algebras. Hence the formalism discussed here is expected to be valid mathematically to describe only type I von Neumann algebras. I shall not elaborate further.

4.4 Gleason's Theorem and the Born Rule

4.4.1 States and Probabilities

In the presentation of the formalism we have not put emphasis on the concept of states, although states are central in the definition of the causality order relation $\preceq$ and of the orthocomplementation $'$. We recall that to each state ψ and to each proposition a is associated the probability $\psi(a)$ for a to be found true on the state ψ. In other word, states are probability measures on the set of propositions, compatible with the causal structure. As already mentioned, the lattice structure of propositions can be formulated from the properties of the states on $\mathcal{L}$.

At that stage we have almost derived the standard mathematical formulation of quantum mechanics. Proposition (yes-no observables) are represented by orthonormal projectors on a Hilbert space $\mathcal{H}$. Projectors on pure states corresponds to projectors on one dimensional subspaces, or rays of $\mathcal{H}$ so the concept of pure states is associated to the vectors of $\mathcal{H}$.

Nevertheless it remains to understand which are the consistent physical states, and what are the rules which determine the probabilities for a proposition a to be true in a state ψ, in particular in a pure state. We remind that the states are in fact characterized by these probability distributions $a \rightarrow \psi(a)$ on $\mathcal{L}$. Thus states must form a convex set of functions $\mathcal{L} \rightarrow [0, 1]$ and by consistency with the OM structure of $\mathcal{L}$ they must satisfy four conditions. These conditions define "quantum probabilities"

Quantum Probabilities

$$(1) \qquad \psi(a) \in [0, 1] \tag{4.48}$$

$$(2) \qquad \psi(\emptyset) = 0, \quad \psi(1) = 1 \tag{4.49}$$

$$(3) \qquad a \neq b \implies \exists \psi \text{ such that } \psi(a) \neq \psi(b) \tag{4.50}$$

$$(4) \qquad a \perp b \implies \quad \psi(a \vee b) = \psi(a) + \psi(b) \tag{4.51}$$

Conditions (1) and (2) are the usual normalization conditions for probabilities. Condition (3) means that observables are distinguishable by their probabilities. Condition (4) is simply the fact that if a and b are orthogonal, they generate a Boolean algebra, and the associated probabilities must satisfy the usual sum rule. These conditions imply in particular that for any state ψ, $\psi(\neg a) = 1 - \psi(a)$, and that if $a \preceq b$, then $\psi(a) \leq \psi(b)$, as we expect.

It remains to understand if and why all states ψ can be represented by density matrices ρ_ψ, and the probabilities for propositions $\mathbf{a}$ given by $\psi(\mathbf{a}) = \text{tr}(\rho_\psi P_\mathbf{a})$, where $P_\mathbf{a}$ is the projector onto the linear-subspace associated to the proposition $\mathbf{a}$. This is a consequence of a very important theorem in operator algebras, Gleason's theorem [50].

4.4.2   Gleason's Theorem

It is easy to see that to obtain quantum probabilities that satisfy the conditions (4.48)–(4.51), it is sufficient to consider atomic propositions, i.e. projections onto one dimensional subspaces (rays) generated by vectors $\vec{e} = |e\rangle$ (pure states) of the Hilbert space $\mathcal{H}$. Indeed, using (4.51), the probabilities for general projectors can be reconstructed (by the usual sum rule) from the probabilities for projections on rays. Denoting since there is no ambiguity for a state ψ the probability for the atomic proposition $\mathbf{e}$ represented by the projection $P_{\vec{e}}$ onto a vector $\vec{e} = |\vec{e}\rangle \in \mathcal{H}$ as

$$\psi(\mathbf{e}) = \psi(P_\mathbf{e}) = \psi(\vec{e}) \qquad P_{\vec{e}} = |\vec{e}\rangle\langle\vec{e}| = \text{projector onto } \vec{e} \tag{4.52}$$

The rules (4.48)–(4.51) reduces for atomic propositions to the conditions.

Quantum Probabilities for Projections on Pure States For any state ψ, the function $\psi(\vec{e})$ considered as a function on the "unit sphere" of the rays over the Hilbert Space $\mathcal{H}$ (the projective space) $\mathcal{S} = \mathcal{H}^*/K^*$ must satisfy

(1) $\psi(\vec{e}) = \psi(\lambda \vec{e})$ for any $\lambda \in K$ such that $|\lambda| = 1$ (4.53)

(2) $0 \le \psi(\vec{e}) \le 1$ (4.54)

(3)   For any complete orthonormal basis of $\mathcal{H}$, $\{\vec{e}_i\}$, one has $\sum_i \psi(\vec{e}_i) = 1$

$$(4.55)$$

Gleason's theorem states the fundamental result that any such function is in one to one correspondence with a density matrix.

Gleason's Theorem If the Hilbert space $\mathcal{H}$ over $K = \mathbb{R}$ or $\mathbb{C}$ is such that

$$\dim(\mathcal{H}) \ge 3 \tag{4.56}$$

then any function ψ over the unit rays of $\mathcal{H}$ that satisfies the three conditions (4.53)–(4.55) is of the form

$$\psi(\vec{e}) = (\vec{e} \cdot \rho_\psi \cdot \vec{e}) = \langle \vec{e}|\rho_\psi|\vec{e}\rangle \tag{4.57}$$

where ρ_ψ is a positive quadratic form (a density matrix) over $\mathcal{H}$ with the expected properties for a density matrix

$$\rho_\psi = \rho_\psi^\dagger \quad , \qquad \rho_\psi \ge 0 \quad , \qquad \mathrm{tr}(\rho_\psi) = 1 \tag{4.58}$$

Reciprocally, any such quadratic form defines a function ψ with the three properties (4.53)–(4.55).

Gleason's theorem is fundamental. As we shall discuss more a bit later, it implies the Born rule. It is also very important when discussing (and excluding a very general and most natural class of) hidden variables theories. So let us discuss it a bit more, without going into the details of the proof.

4.4.3 Principle of the Proof

The theorem is remarkable since there are non conditions on the regularity or measurability of the function ψ. In the original derivation by Gleason [50] he considers real "frame functions" f of weight W over $\mathcal{H}^* = \mathcal{H}\backslash\{0\}$ such that

(1) $f(\vec{e}) = f(\lambda\vec{e})$ for any $\lambda \neq 0 \in K$ (4.59)

(2) f is bounded   (4.60)

(3)   For any complete orthonormal basis of $\mathcal{H}$, $\{\vec{e}_i\}$, $\sum_i f(\vec{e}_i) = W = $ constant

(4.61)

and proves that such a function must be of the form (4.57)

$$f(\vec{e}) = (\vec{e} \cdot Q \cdot \vec{e}) \quad , \qquad Q \quad \text{quadratic form such that} \quad \text{tr}(Q) = W \qquad (4.62)$$

It is easy to see that this is equivalent to the theorem as stated above, since one can add constants and rescale the functions f to go from (4.59)–(4.61) to (4.53)–(4.55). The original proof goes into three steps

1. Real Hilbert space, $\dim(\mathcal{H}) = 3$ and f a continuous frame function $\implies$ the theorem

    This is the easiest part, involving some group theory. Any frame function f is a real function on the unit two dimensional sphere S_2 and if continuous it is square summable and can be decomposed into spherical harmonics

$$f(\vec{n}) = \sum_{l,m} f_{l,m} Y_l^m(\theta, \varphi) \qquad (4.63)$$

The theorem amounts to show that if f is a frame function of weight $W = 0$, then only the $l = 2$ components of this decomposition (4.63) are non zero. Some representation theory (for the $SO(3)$ rotation group) is enough. Any orthonormal (oriented) basis $(\vec{e}_1, \vec{e}_2, \vec{e}_3)$ of $\mathbb{R}^3$ is obtained by applying a rotation R to the basis $(\vec{e}_x, \vec{e}_y, \vec{e}_z)$. Thus one can write

$$f(\vec{n}_1) + f(\vec{n}_2) + f(\vec{n}_3) = \sum_l \sum_{m,m'} f_{l,m} \, D_{m,m'}^{(l)}(R) \, V_{m'}^{(l)} \qquad (4.64)$$

with the $D_{m,m'}^{(l)}(R)$ the Wigner D matrix for the rotation R, and the $V_{m'}^{(l)}$ the components of the vectors $\vec{V}^{(l)}$ in the spin l representation of $SO(3)$, with components

$$\vec{V}^{(l)} = \{V_m^{(l)}\} \quad , \qquad V_m^{(l)} = Y_{l,m}(0,0) + Y_{l,m}(\pi/2,0) + Y_{l,m}(\pi/2,\pi/2) \qquad (4.65)$$

If f is a frame function of weight $W = 0$, the l.h.s. of (4.64) is zero for any $R \in SO(3)$. This implies that for a given l, the coefficients f_{lm} must vanish if the vector $\vec{V}^{(l)} \neq 0$, but are undetermined if $\vec{V}^{(l)} = 0$. An explicit calculation shows that indeed

$$\vec{V}^{(l)} \begin{cases} \neq 0 & \text{if } l \neq 2, \\ = 0 & \text{if } l = 2. \end{cases} \tag{4.66}$$

This establishes the theorem in case (1).

2. Real Hilbert space, $\dim(\mathcal{H}) = 3$ and f any frame function $\implies f$ continuous.

    This is the most non-trivial part: assuming that the function is bounded, the constraint (4.61) is enough to imply that the function is continuous! It involves a clever use of spherical geometry and of the frame identity $\sum_{i=1,2,3} f(\vec{e}_i) = W$. The basic idea is to start from the fact that since f is bounded, it has a lower bound f_{min} which can be set to 0. Then for any $\epsilon > 0$, take a vector $\vec{n}_0$ on the sphere such that $|f(\vec{n}_0) - f_{min}| < \epsilon$. It is possible to show that there is a neighbourhood $\mathcal{O}$ of $\vec{n}_0$ such that $|f(\vec{n}_1) - f(\vec{n}_2)| < C \epsilon$ for any $\vec{n}_1$ and $\vec{n}_2 \in \mathcal{O}$. C is a universal constant. It follows that the function f is continuous at its minimum! Then it is possible, using rotations to show that the function f is continuous at any points on the sphere.

3. Generalize to $\dim(\mathcal{H}) > 3$ and to complex Hilbert spaces.

    This last part is more standard and more algebraic. Any frame function $f(\vec{n})$ defined on unit vectors $\vec{n}$ such that $\|\vec{n}\| = 1$ may be extended to a quadratic function over vectors $f(\vec{v}) = \|\vec{v}\|^2 f(\vec{v}/\|\vec{v}\|)$.

    For a real Hilbert space with dimension $d > 3$, the points (1) and (2) imply that the restriction of a frame function $f(\vec{n})$ to any three dimensional subspace is a quadratic form $f(\vec{v}) = (\vec{v} \cdot Q \cdot \vec{v})$. A simple and classical theorem by Jordan and von Neumann shows that this is enough to define a global real quadratic form Q on the whole Hilbert space $\mathcal{H}$ through the identity $2(\vec{x} \cdot Q \cdot \vec{y}) = f(\vec{x} + \vec{y}) - f(\vec{x} - \vec{y})$.

    For complex Hilbert spaces, the derivation is a bit more subtle. One can first apply the already obtained results to the restriction of frame functions over real submanifolds of $\mathcal{H}$ (real submanifolds are real subspaces of $\mathcal{H}$ such that $(\vec{x} \cdot \vec{y})$ is always real). One can then extend the obtained real quadratic form over the real submanifolds to a complex quadratic form on $\mathcal{H}$.

4.4.4 The Born Rule

The Born rule is a simple consequence of Gleason theorem. Indeed, any state (in the general sense of statistical state) corresponds to a positive quadratic form (a density matrix) ρ. Given a minimal atomic proposition, which corresponds to a projector $P = |\vec{a}\rangle\langle\vec{a}|$ onto the ray corresponding to a single vector (pure state) $|\vec{a}\rangle$, the probability p for P of being true is

$$p = \langle P \rangle = \text{tr}(\rho P) = \langle \vec{a} | \rho | \vec{a} \rangle \tag{4.67}$$

The space of states $\mathcal{E}$ is thus the space of (symmetric) positive density matrices with unit trace

$$\text{space of states} = \mathcal{E} = \{\rho : \rho = \rho^\dagger, \rho \geq 0, \text{tr}(\rho) = 1\} \tag{4.68}$$

It is a convex set. Its extremal points, which cannot be written as a linear combination of two different states, are the pure states of the system, and are the density matrices of rank one, i. e. the density matrices which are themselves projectors onto a vector $|\psi\rangle$ of the Hilbert space.

$$\rho = \text{pure state} \quad \Longrightarrow \quad \rho = |\psi\rangle\langle\psi| \quad, \quad \|\psi\| = 1 \tag{4.69}$$

One thus derives the well known fact that the pure states are in one to one correspondence with the vectors (well... the rays) of the Hilbert space $\mathcal{H}$ that was first introduced from the basic observables of the theory, the elementary atomic propositions (the projectors P). Similarly, one recovers the simplest version of the Born rule: the probability to "observe" a pure state $|\varphi\rangle$ into the (non orthogonal) pure state $|\psi\rangle$ (to "project" $|\varphi\rangle$ onto $|\psi\rangle$) is the square of the norm of the scalar product

$$p(\varphi \to \psi) = |\langle\varphi|\psi\rangle|^2 \tag{4.70}$$

4.4.5 Physical Observables

One can easily reconstruct the set of all physical observables, and the whole algebra of observables $\mathcal{A}$ of the system. I present the line of the argument, without any attempt of mathematical rigor.

Any ideal physical measurement of some observable O consists in fact in taking a family of mutually orthogonal propositions a_i, i.e. of commuting symmetric projectors P_i on $\mathcal{H}$ such that

$$P_i^2 = P_i, \quad P_i = P_i^*, \quad P_i P_j = P_j P_i = 0 \text{ if } i \neq j \tag{4.71}$$

performing all the tests (the order is unimportant since the projectors commute) and assigning a *real number* o_i to the result of the measurement (the value of the observable O) if a_i is found true (this occurs for at most one a_i) and zero otherwise. In fact one should take an appropriate limit when the number of a_i goes to infinity, but I shall not discuss these important points of mathematical consistency. If you think about it, this is true for any imaginable measurement (position, speed, spin, energy, etc.). The resulting physical observable O is thus associated to the symmetric operator

$$O = \sum_i o_i P_i \tag{4.72}$$

This amounts to the spectral decomposition of symmetric operators in the theory of algebras of operators.

Consider a system in a general state given by the density matrix ρ. From the general rules of quantum probabilities, the probability to find the value o_i for the measurement of the observable O is simply the sum of the probabilities to find the system in a eigenstate of O of eigenvalue o_i, that is

$$p(O \to o_i) = \mathrm{tr}(P_i \rho) \tag{4.73}$$

and for a pure state $|\varphi\rangle$ it is simply

$$\langle \varphi | P_i | \varphi \rangle = |\langle \varphi | \varphi_i \rangle|^2 \quad , \qquad |\varphi_i\rangle = \frac{1}{\| P_i | \varphi \rangle \|} \, P_i | \varphi \rangle \tag{4.74}$$

Again the Born rule! The expectation value for the result of the measurement of O in a pure state ψ_a is obviously

$$\mathbb{E}[O; \psi] = \langle O \rangle_\psi = \sum_i o_i \, p(O \to o_i) = \sum_i o_i \langle \psi | P_i | \psi \rangle = \langle \psi | O | \psi \rangle \tag{4.75}$$

This is the standard expression for expectation values of physical observables as diagonal matrix elements of the corresponding operators. Finally for general (mixed) states one has obviously

$$\mathbb{E}[O; \rho] = \langle O \rangle_\rho = \sum_i o_i \, p(O \to o_i) = \sum_i o_i \, \mathrm{tr}(P_i \rho) = \mathrm{tr}(O \rho) \tag{4.76}$$

We have seen that the pure states generate by convex combinations the convex set $\mathcal{E}$ of all (mixed) states ψ of the system. Similarly the symmetric operators $O = O^\dagger$ generates (by operator multiplication and linear combinations) a C*-algebra $\mathcal{A}$ of bounded operators $\mathcal{B}(\mathcal{H})$ on the Hilbert space $\mathcal{H}$. States are normalized positive linear forms on $\mathcal{A}$ and we are back to the standard algebraic formulation of quantum physics. The physical observables generates an algebra of operators, hence an abstract algebra of observables, as assumed in the algebraic formalism. We refer to the section about the algebraic formalism for the arguments for preferring complex Hilbert spaces to real or quaternionic ones.

4.5 Discussion

This chapter was a sketchy and partial introduction to the quantum logic approach for the formulation of the principle of quantum mechanics. I hope to have shown its relation with the algebraic formulation. It relies on the concepts of states and of observables as the algebraic formulation. However the observables are limited to the physical subset of yes/no proposition, corresponding to ideal projective measurements, without assuming a priori some algebraic structure between non-compatible propositions (non-commuting observables in the algebraic framework). I explained how the minimal set of axioms on these propositions and their actions on states, used in the quantum logic approach, is related to the physical concepts of causality, reversibility and separability/locality. The canonical algebraic structure of quantum mechanics comes out from the symmetries of the "logical structure" of the lattice of propositions. The propositions corresponding to ideal projective measurements are realized on orthogonal projections on a (possibly generalized) Hilbert space. Probabilities/states are given by quadratic forms, and the Born rule follows from the logical structure of quantum probabilities through Gleason's theorem.

The quantum logic approach is of course not completely foolproof. We have seen that the issue of the possible division ring K is not completely settled. The strong assumptions of atomicity and covering are essential, but somehow restrictive compared to the algebraic approach (type II and III von Neumann algebras). It is sometime stated that it cannot treat properly the case of a system composed of two subsystems since there is no concept of "tensorial product" of two OM-AC lattices as there is for Hilbert spaces and operator algebras. Note however that one should in general always think about multipartite systems as parts of a bigger system, not the opposite! Even in the algebraic formulation it is not known in the infinite dimensional case if two commuting subalgebras $\mathcal{A}_1$ and $\mathcal{A}_2$ of a bigger C*-algebra $\mathcal{A}$ always correspond to the decomposition of the Hilbert space $\mathcal{H}$ into a tensor product of two subspaces $\mathcal{H}_1$ and $\mathcal{H}_2$ (this is known as the Tsirelson conjecture).

Chapter 5
Information, Correlations, and More

5.1 Quantum Information Formulations

Quantum information science has experienced enormous developments during the last 30 years. I do not cover this wide and fascinating field in these notes, but shall only discuss briefly some relations with the question of the formalism. Indeed quantum information theory led to new points of view and to new uses and applications of quantum theory. This renewal is considered by some authors as a real change of paradigm, and referred to as "the second quantum revolution".

From what I know, the interest in the relations between Information Theory and Quantum Physics started really in the 1970s, from the confluence of several questions and new results. Let me quote a few.

- This period experienced a better understanding of the relations and of the conflicts between General Relativity and Quantum Mechanics: the theoretical discovery of the Bekenstein-Hawking quantum entropy for black holes, the black hole evaporation (information) paradox, the more general Unruh effect and quantum thermodynamical aspects of gravity and of events horizons (with more recently many developments in quantum gravity and string theories, such as "Holographic gravity", "Entropic Gravity", etc.).
- The ongoing discussions on the various interpretations of the quantum formalism, the meaning of quantum measurement processes, and whether a quantum state represent the "reality", or some "element of reality" on a quantum system, or simply the observer's information on the quantum system, experienced a revival through the development of the concept of decoherence.
- The 1970s saw of course the theoretical and experimental developments of quantum computing. A standard reference is the book by Nielsen and Chuang [82]. This field started from the realization that quantum entanglement and quantum correlations can be used as a resource for performing calculations and

© Springer International Publishing Switzerland 2015 105
F. David, *The Formalisms of Quantum Mechanics*, Lecture Notes in Physics 893,
DOI 10.1007/978-3-319-10539-0_5

the transmission of information in a more efficient way than when using classical correlations with classical channels.

- This led for instance to the famous "It from Bit" idea (or aphorism) of J. A. Wheeler (see e.g. in [115]) and others (see for instance the book by Deutsch [36], or talks by Fuchs [44,45]). Roughly speaking this amounts to reverse the famous statement of Laudauer "Information is Physics" into "Physics is Information", and to state that Information is the good starting point to understand the nature of the physical world and of the physical laws.

This point of view that Quantum Physics has to be considered from the point of view of Information has been developed and advocated by several authors in the area of quantum gravity and quantum cosmology. Here I shall just mention some old or recent attempts to use this point of view to discuss the formalism of "standard" quantum physics, not taking into account the issues of quantum gravity.

In the quantum-information-inspired approaches a basic concept is that of "device", or "operation", which represents the most general manipulation on a quantum system. In a very oversimplified presentation,[1] such a device is a "black box" with both a quantum input system A and quantum output system B, and with a set I of classical settings $i \in I$ and a set O of classical responses $o \in O$. The outputs do not need to be yes/no answers to a set of compatible quantum observables (orthonormal basis in the standard formalism) but may be more general (for instance associated to a POVM). The input and output systems A and B may be different, and may be multipartite systems, e.g. may consist in collections of independent subsystems $A = \bigcup_\alpha A_\alpha$, $B = \bigcup_\alpha B_\beta$.

This general concept of device encompass the standard concepts of *state* and of *effect*. A *state* corresponds to the preparation of a quantum system S in a definite state; there is no input $A = \emptyset$, the setting i specify the state, there is no response, and $B = S$ is the system. An *effect*, corresponds to a destructive measurement on a quantum system S; the input $A = S$ is the system, there are no output $B = \emptyset$, no settings i, and the response set O is the set of possible output measurements o. This concept of device contains also the general concept of a *quantum channel*; then $A = B$, there are no settings or responses (Fig. 5.1).

Probabilities $p(i|o)$ are associated to the combination of a state and an effect, this correspond to the standard concept of probability of observing some outcome o when making a measurement on a quantum state (prepared according to i) (Fig. 5.2).

General information processing quantum devices are constructed by building causal circuits out of these devices used as building blocks, thus constructing complicated apparatus out of simple ones. An information theoretic formalism is obtained by choosing axioms on the properties of such devices (states and effects) and operational rules to combine these devices and circuits and the associated probabilities, thus obtaining for instance what is called in [23] an "operational probabilistic theory". This framework is also called "generalized probabilistic

---

[1] Slightly more general than in some presentations.

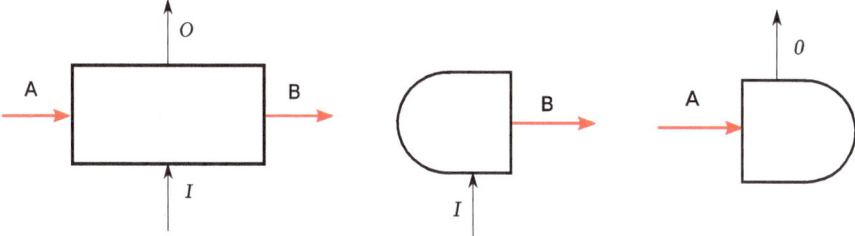

Fig. 5.1 A general device, a state and an effect

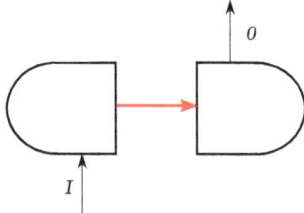

Fig. 5.2 Probabilities are associated to a couple state-effect

theories", where "preparations", "transformations" and "measurements" replace the concepts of state, channel and effect.

This kind of approach is mainly considered for finite dimensional theories (which in the quantum case correspond to finite dimensional Hilbert spaces), both because it seems problematic to formulate them properly in the infinite dimensional case, and because systems with a finite number of distinct orthogonal pure states are those that are usually considered in quantum information science.

This approach leads to a pictorial formulation of quantum information processing. It shares similarities with the "quantum pictorialism" logic formalism, rather based on category theory, and presented for instance in [26].

It can also be viewed as an operational and informational extension of the older "convex set approach" (developed notably by G. Ludwig, see [76] and [8, 15] for details). This later approach, which is also related to the quantum logic approach, puts more emphasis on the concept of states than on the concept of observables in quantum mechanics.

I shall not discuss much further these quantum information-theoretic approaches, since I did not know much about information theory and quantum information. Let me just highlight the recent attempts of Hardy [58, 59] and those of Chiribella et al. [22, 23] (see [20] for a short presentation of this last formulation). Another proposal is that of Masanes and Müller [79].

In [23] the standard complex Hilbert space formalism of QM is derived from six informational principles: Causality, Perfect Distinguishability, Ideal Compression, Local Distinguishability, Pure Conditioning and Purification Principle. The first two principles are not very different from the principles of other formulations (causality is defined in a standard sense, and distinguishability is related to the concept of differentiating states by measurements). The third one is related to the existence of reversible maximally efficient compression schemes for states. The four and the fifth are about the properties of bipartite states and for instance the possibility to performing local tomography and the effect of separate "atomic" measurements on such states. The last one, "purification principle", distinguishes quantum mechanics from classical mechanics, and states that any mixed state of some system $\mathcal{S}$ may be obtained from a pure state of a larger composite system $\mathcal{S} + \mathcal{S}'$. See [20] for a discussion of the relation of this purification principle with the discussions of the "Heisenberg cut" between the system measured and the measurement device (see Heisenberg's 1935 reply to EPR in [31]). Of course this is related to previous discussions by von Neumann in [105].

In [79] a set of five informational axioms on the properties of physics states allow to derive the formalism of quantum (and of classical) mechanics. This approach shares elements with [58, 59], but the fifth axiom is simpler. Again the existence of reversible continuous transformations between states (axiom 4') is a very important feature to obtain by geometrical argument the standard Hilbert space structure for (finite dimensional) quantum mechanics.

5.2   Quantum Correlations

The world of quantum correlations is richer, more subtle and more interesting than the world of classical correlations. Most of the puzzling features and apparent paradoxes of quantum physics come from the properties of these correlations, especially for entangled states. Entanglement is probably *the* distinctive feature of quantum mechanics, and is a consequence of the superposition principle when considering quantum states for composite systems. Here I discuss briefly some basic aspects. Entanglement describes the particular quantum correlations between two quantum systems which (for instance after some interactions) are in a non separable pure state, so that each of them considered separately, is not in a pure state any more. Without going into history, let me remind that if the terminology "entanglement" ("Verschränkung") was introduced in the quantum context by E. Schrödinger in 1935 (when discussing the famous EPR paper). However the mathematical concept is older and goes back to the modern formulation of quantum mechanics. For instance, some peculiar features of entanglement and its consequences have been discussed already around the 1930 in relation with the theory of quantum measurement by Heisenberg, von Neumann, Mott, etc. Examples of interesting entangled many particles states are provided by the Stater determinant for many fermion states, by the famous Bethe ansatz for the ground state of the spin 1/2 quantum chain, etc.

5.2.1 Entropic Inequalities

von Neumann Entropy The difference between classical and quantum correlations is already visible when considering the properties of the von Neumann entropy of states of composite systems. Remember that the von Neumann entropy of a mixed state of a system A, given by a density matrix ρ_A, is given by

$$S(\rho_A) = -\mathrm{tr}(\rho_A \log \rho_A) \tag{5.1}$$

In quantum statistical physics, the log is usually the natural logarithm

$$\log = \log_e = \ln \tag{5.2}$$

while in quantum information, the log is taken to be the binary logarithm

$$\log = \log_2 \tag{5.3}$$

The entropy measures the amount of "lack of information" that we have on the state of the system. But in quantum physics, at variance with classical physics, one must be very careful about the meaning of "lack of information", since one cannot speak about the precise state of a system before making measurements. So the entropy could (and should) rather be viewed as a measure of the number of independent measurements we can make on the system before having extracted all the information, i.e. the amount of information we can extract of the system. It can be shown also that the entropy give the maximum information capacity of a quantum channel that we can build out of the system. See [82] for a good introduction to quantum information and in particular on entropy viewed from the information theory point of view.

When no ambiguity exists on the state ρ_A of the system A, I shall use the notations

$$S_A = S(A) = S(\rho_A) \tag{5.4}$$

The von Neumann entropy shares many properties of the classical entropy. It has the same convexity properties

$$S[\lambda\rho + (1-\lambda)\rho'] \geq \lambda S[\rho] + (1-\lambda)S[\rho'] \quad , \qquad 0 \leq \lambda \leq 1 \tag{5.5}$$

It is minimal $S = 0$ for systems in a pure state and maximal for systems in a equipartition state $S = \log(N)$ if $\rho = \frac{1}{N}\mathbf{1}_N$. It is extensive for systems in separate states.

Relative Entropy The relative entropy (of a state ρ w.r.t. another state σ for the same system) is defined as in classical statistics (Kullback-Leibler entropy) as

$$S(\rho\|\sigma) = \mathrm{tr}(\rho \log \rho) - \mathrm{tr}(\rho \log \sigma) \tag{5.6}$$

with the same convexity properties.

The differences with the classical entropy arise for composite systems. For such a system AB, composed of two subsystems A and B, a general mixed state is given by a density matrix ρ_{AB} on $\mathcal{H}_{AB} = \mathcal{H}_A \otimes \mathcal{H}_B$. The reduced density matrices for A and B are

$$\rho_A = \mathrm{tr}_B(\rho_{AB}) \quad , \qquad \rho_B = \mathrm{tr}_A(\rho_{AB}) \tag{5.7}$$

This corresponds to the notion of marginal distribution w.r.t. A and B of the general probability distribution of states for AB in classical statistics. Now if one considers

$$S(AB) = -\mathrm{tr}(\rho_{AB} \log \rho_{AB}), \quad S(A) = -\mathrm{tr}(\rho_A \log \rho_A), \quad S(B) = -\mathrm{tr}(\rho_B \log \rho_B) \tag{5.8}$$

one has the following definitions.

Conditional Entropy The *conditional entropy* S(A|B) (the entropy of A conditional to B in the composite system AB) is

$$S(A|B) = S(AB) - S(B) \tag{5.9}$$

The conditional entropy $S(A|B)$ corresponds to the remaining uncertainty (lack of information) on A if B is known.

Mutual Information The *mutual information* (shared by A and B in the composite system AB)

$$S(A : B) = S(A) + S(B) - S(AB) \tag{5.10}$$

Subadditivity The entropy satisfies the general inequalities (triangular inequalities)

$$|S(A) - S(B)| \le S(AB) \le S(A) + S(B) \tag{5.11}$$

The rightmost inequality $S(AB) \le S(A) + S(B)$ is already valid for classical systems, but the leftmost is quantum. Indeed for classical systems the classical entropy H_{cl} satisfy only the much stronger lower bound

$$\max(H_{\mathrm{cl}}(A), H_{\mathrm{cl}}(B)) \le H_{\mathrm{cl}}(AB) \tag{5.12}$$

Subadditivity implies that if AB is in a pure entangled state, $S(A) = S(B)$. It also implies that the mutual information in a bipartite system is always positive

$$S(A : B) \geq 0 \qquad (5.13)$$

In the classical case the conditional entropy is always positive $H_{cl}(A|B) \geq 0$. In the quantum case the conditional entropy may be negative $S(A|B) < 0$ if the entanglement between A and B is large enough. This is a crucial feature of quantum mechanics. If $S(A|B) < 0$ it means that A and B share information resources (through entanglement) which get lost if one gets information on B only (through a measurement on B for instance).

Strong Subadditivity Let us consider a tripartite systems ABC. The entropy satisfies another very interesting inequality

$$S(A) + S(B) \leq S(AC) + S(BC) \qquad (5.14)$$

It is equivalent to (this is the usual form)

$$S(ABC) + S(C) \leq S(AC) + S(BC) \qquad (5.15)$$

Note that (5.14) is also true for the classical entropy, but then for simple reasons. In the quantum case it is a non trivial inequality.

The strong subadditivity inequality implies the triangle inequality for tripartite systems

$$S(AC) \leq S(AB) + S(BC) \qquad (5.16)$$

so the entropic inequalities can be represented graphically as in Fig. 5.3

The strong subadditivity inequality has important consequences for conditional entropy and mutual information (see [82]). Consider a tripartite composite system ABC. It implies for instance

$$S(C|A) + S(C|B) \geq 0 \qquad (5.17)$$

and

$$S(A|BC) \leq S(A|B) \qquad (5.18)$$

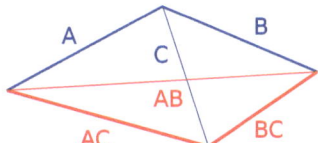

Fig. 5.3 Entropic inequalities: the length of the line "X" is the von Neumann entropy $S(X)$. The tetrahedron has to be "oblate", the sum AC+BC (*fat red lines*) is always $\geq$ the sum A+B (*fat blue lines*)

which means that conditioning A to a part of the external subsystem (here C inside BC) increase the information we have on the system (here A). One has also for the mutual information

$$S(A : B) \leq S(A : BC) \tag{5.19}$$

This means that discarding a part of a multipartite quantum system (here C) increases the mutual information (here between A and the rest of the system). This last inequality is very important. It implies for instance that if one has a composite system AB, performing some quantum operation on B without touching to A cannot increase the mutual information between A and the rest of the system.

Let us mention other subadditivity inequalities for tri- or quadri-partite systems.

$$S(AB|CD) \leq S(A|C) + S(B|D) \tag{5.20}$$

$$S(AB|C) \leq S(A|C) + S(B|C) \tag{5.21}$$

$$S(A|BC) \leq S(A|B) + S(A|C) \tag{5.22}$$

5.2.2 Bipartite Correlations

The specific properties of quantum correlations between two causally separated systems are known to disagree with what one would expect from a "classical picture" of quantum theory, where the quantum probabilistics features come just from some lack of knowledge of underlying "elements of reality". I shall come back later on the very serious problems with the "hidden variables" formulations of quantum mechanics. But let us discuss already some of the properties of these quantum correlations in the simple case of a bipartite system.

I shall discuss briefly one famous and important result: the Tsirelson bound. The general context is that of the discussion of non-locality issues and of Bell's [13] and CHSH inequalities [25] in bipartite systems. However, since these last inequalities are more of relevance when discussing hidden variables models, I postpone their discussion to the next Sect. 5.3.

This presentation is standard and simply taken from [71].

5.2.2.1 The Tsirelson Bound

The Two Spin System Consider a simple bipartite system consisting of two spins 1/2, or q-bits 1 and 2. If two observers (Alice $\mathcal{A}$ and Bob $\mathcal{B}$) make independent measurements of respectively the value of the spin 1 along some direction $\vec{n}_1$ (a unit vector in 3D space) and of the spin 2 along $\vec{n}_2$, at each measurement they get results (with a correct normalization) +1 or −1. Now let us compare the results of four experiments, depending whether $\mathcal{A}$ choose to measure the spin 1 along a first direction $\vec{a}$ or a second direction $\vec{a}'$, and wether $\mathcal{B}$ chose (independently) to measure the spin 2 along a first direction $\vec{b}$ or a second direction $\vec{b}'$. Let us call the corresponding observables **A**, **A′**, **B**, **B′**, and by extension the results of the corresponding measurements in a single experiment A and A' for the first spin, B and B' for the second spin.

$$\text{spin 1 along } \vec{a} \quad \rightarrow \quad A = \pm 1 \quad ; \quad \text{spin 1 along } \vec{a}' \quad \rightarrow \quad A' = \pm 1 \quad (5.23)$$

$$\text{spin 2 along } \vec{b} \quad \rightarrow \quad B = \pm 1 \quad ; \quad \text{spin 2 along } \vec{b}' \quad \rightarrow \quad B' = \pm 1 \tag{5.24}$$

Now consider the following combination M of products of observables, hence of products of results of experiments

$$M = AB - AB' + A'B + A'B' \tag{5.25}$$

and consider the expectation value $\langle M \rangle_\psi$ of M for a given quantum state $|\psi\rangle$ of the two spins system. In practice this means that we prepare the spins in state $|\psi\rangle$, chose randomly (with equal probabilities) one of the four observables, and to test locality $\mathcal{A}$ and $\mathcal{B}$ may be causally deconnected, and choose independently (with equal probabilities) one of their own two observables, i.e. spin directions. Then they make their measurements. The experiment is repeated a large number of time and the right average combination M of the results of the measurements is calculated afterwards.

A simple explicit calculation shows the following inequality, known as the Tsirelson bound [24]

Tsirelson Bound For any state and any choice orientations $\vec{a}$, $\vec{a}'$, $\vec{b}$ and $\vec{b}'$, one has

$$|\langle M \rangle| \leq 2\sqrt{2} \tag{5.26}$$

while, as discussed later, "classically", i.e. for theories where the correlations are described by contextually-local hidden variables attached to each subsystem, one has the famous Bell-CHSH bound

$$\langle |M| \rangle_{\text{"classical"}} \leq 2. \tag{5.27}$$

The Tsirelson bound is saturated if the state $|\psi\rangle$ for the two spin is the singlet

$$|\psi\rangle = |\text{singlet}\rangle = \frac{1}{\sqrt{2}} \left(|\uparrow\rangle \otimes |\downarrow\rangle - |\downarrow\rangle \otimes |\uparrow\rangle \right) \tag{5.28}$$

and the directions for $\vec{a}$, $\vec{a}'$, $\vec{b}$ and $\vec{b}'$ are coplanar, and such that $\vec{a} \perp \vec{a}'$, $\vec{b} \perp \vec{b}'$, and the angle between $\vec{a}$ and $\vec{b}$ is $\pi/4$, as depicted on Fig. 5.4.

Fig. 5.4 Spin directions for
saturating the Tsirelson
bound and maximal violation
of the Bell-CHSH inequality

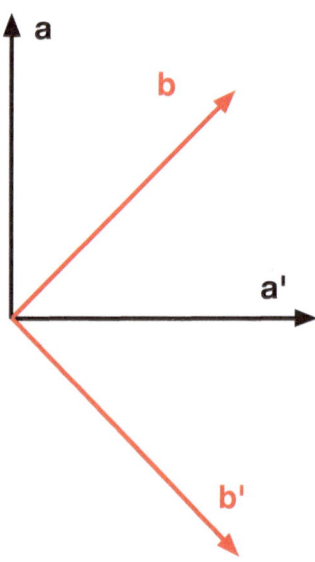

5.2.2.2 Popescu-Rohrlich Boxes

Beyond the Tsirelson Bound ? Interesting questions arise when one consider what
could happen if there are "super-strong correlations" between the two spins (or in
general between two subsystems) that violate the Tsirelson bound. Indeed, the only
mathematical bound on M for general correlations is obviously $|\langle M \rangle| \leq 4$. Such
hypothetical systems are considers in the theory of quantum information and are
denoted Popescu-Rohrlich boxes [89]. With the notations of the previously consid-
ered 2 spin system, BR-boxes consist in a collection of probabilities $P(A, B|a, b)$
for the outputs A and B of the two subsystems, the input or settings a and b being
fixed. The (a, b) correspond to the settings I and the (A, B) to the outputs O of
Fig. 5.1 of the quantum information section. In our case we can take for the first spin

$$a = 1 \; \rightarrow \; \text{chose orientation } \vec{a} \quad , \qquad a = -1 \; \rightarrow \; \text{chose orientation } \vec{a}' \quad (5.29)$$

and for the second spin

$$b = 1 \; \rightarrow \; \text{chose orientation } \vec{b} \quad , \qquad b = -1 \; \rightarrow \; \text{chose orientation } \vec{b}' \quad (5.30)$$

The possible outputs being always $A = \pm 1$ and $B = \pm 1$.

The fact that the $P(A, B|a, b)$ are probabilities means that

$$0 \leq P(A, B|a, b) \leq 1 \quad , \qquad \sum_{A,B} P(A, B|a, b) = 1 \quad \text{for} \quad a, b \text{ fixed} \quad (5.31)$$

Non Signalling If the settings a and b and the outputs A and B are relative to two causally separated parts of the system, corresponding to manipulations by two independent agents (Alice and Bob), enforcing causality means that Bob cannot guess which setting (a or a') Alice has chosen from his choice of setting (b and b') and his output (B or B), without knowing Alices' output A. The same holds for Alice with respect to Bob. This requirement is enforced by the non-signaling conditions (Fig. 5.5)

$$\sum_A P(A, B|a, b) = \sum_A P(A, B|a', b) \qquad (5.32)$$

$$\sum_B P(A, B|a, b) = \sum_B P(A, B|a, b') \qquad (5.33)$$

A remarkable fact is that there are choices of probabilities which respect the non-signaling condition (hence causality) but violate the Tsirelson bound and even saturate the absolute bound $|\langle M \rangle| = 4$. Such hypothetical devices would allow to use "super-strong correlations" (also dubbed "super-quantum correlations") to manipulate and transmit information in a more efficient way than with quantum systems (in the standard way of quantum information protocols, by sharing some initially prepared bipartite quantum system and exchanges of classical information) [18, 84, 102]. However, besides these very intriguing features of "trivial communication complexity", such devices are problematic. In particular it seems that no interesting dynamics can be defined on such systems [54].

Before concluding the discussion on the nature of quantum correlations, it is of course important to discuss their "non-local" nature and some of the issues related to the EPR-paradox and the attempts to explain theses correlations by hidden variables models. I discuss some of theses questions, in particular Bell inequalities, in the next Sect. 5.3. A summary will be made afterwards in Sect. 5.4.

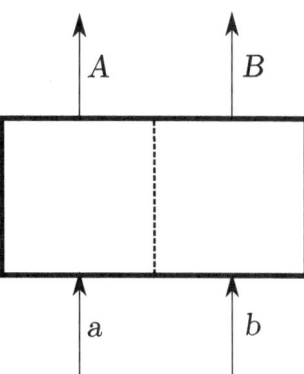

Fig. 5.5 A Popescu-Rohrlich box

5.3 Hidden Variables, Contextuality and Local Realism

5.3.1 Hidden Variables and "Elements of Reality"

In this section I discuss briefly some features of quantum correlations which are important when discussing the possibility that the quantum probabilities may still have, to some extent, a "classical interpretation" by reflecting our ignorance of inaccessible "sub-quantum" degrees of freedom or "elements of reality" of quantum systems, which could behave in a more classical and deterministic way. In particular a question is: which general constraints on the properties of such degrees of freedom are enforced by quantum mechanics?

This is the general idea of the "hidden variables" program and of the search of explicit hidden variable models. These ideas go back to the birth of quantum mechanics, and were in particular proposed by L. de Broglie in his first "pilot wave model", but they were dismissed by most physicists after the famous 1927 Solvay Congress and the tremendous advances of quantum physics in the 1930s. Despite the discussions that followed the EPR paper of 1935, hidden variables models underwent a revival only in the 1960s, from the works of D. Bohm, and mostly from the work of J. Bell, and the subsequent experimental developments that started in the 1970s.

Let me state first that I am not going to review in details or advocate hidden variable models. This section will illustrate the very strong constraints that the quantum formalism enforces on the basic idea. For more in-depth analysis and some incisive criticism see for instance [86] and [96].

In most discussions about realism and locality, see for instance [86] and [71], the Bell's inequality and the issues of non-locality are discussed before the question of "contextuality" (the concept will be explained below). I choose in these notes to discuss first contextuality, since I think it is a more general issue that puts the other problems in perspective, and I shall discuss the Bell's inequalities and some hidden variable ideas after.

The basic idea of hidden variables is that the quantum states (the $|\psi\rangle$'s) of a quantum system $\mathcal{S}$ could be described by some ensembles of (partially or totally) hidden variables $\mathfrak{v}$. These hidden variables $\mathfrak{v}$ belong to some set (probability space) $\mathfrak{V}$, with some unknown statistics and possibly some unknown dynamics. Each $\mathfrak{v}$ (an element of the set $\mathfrak{V}$) represents an instance of a possibly infinite collection of these more fundamental sub-quantum variables. The specific outcome a (a real number) of a measurement operation of a physical observable A is supposed to be determined by the actual hidden variable $\mathfrak{v}$.

$$\text{observable } A \: + \: h.v. \: \mathfrak{v} \: \xrightarrow{\text{measurement}} \: \text{outcome } a = \text{function}(A, \mathfrak{v}) \qquad (5.34)$$

In these schemes, the quantum indeterminism is not fundamental. It is assumed to come from our lack of knowledge about the exact state of the hidden variables. In other word, the pure quantum states $|\psi\rangle$ of the system should correspond to some classical probability distributions $p_\psi(\mathfrak{v})$ on $\mathcal{V}$. Note that the outcome of a measurement may depend not only on the choice of the observable that is measured, but also on the choice and the details of the measurement apparatus. This is usually denoted the "context" of the measurement. One might expect that, if the idea of hidden variable makes sense, the influence of the (partially unknown) context may be taken into account through the hidden variables themselves. This notion of "context of a measurement" and of "contextuality" is a very important one (it is basically the whole point in the discussion of hidden variables) and it will be discussed more in the next sections. Finally the measurement process should be amenable to a description in this framework, taking into effect the possible (deterministic) back reaction of a measurement operation on the hidden variables $\mathfrak{v}$.

A very simple class of hidden variable models has been discussed by J. von Neumann in his 1932 book [105, 106]. The hidden variables (the element of reality) are assumed to be in one to one correspondence with the possible outcomes a of all the observables A of the system. $\mathfrak{v} = \{$outcomes of observables$\}$. Denoting by $f(A, \mathfrak{v})$ the well determined outcome a of a measurement of the observable A (one speaks of "dispersion free" variables) this function must then satisfy the addition law

$$C = A + B \quad \Longrightarrow \quad c = a + b \quad \text{i.e} \quad f(C, \mathfrak{v}) = f(A, \mathfrak{v}) + f(B, \mathfrak{v})$$
$$(5.35)$$

As argued in [105, 106], this is clearly inconsistent with quantum mechanics. Indeed if A and B do not commute, the possible outcomes of C (the eigenvalues of the operator C) are not in general sums of outcomes of A and B (sums of eigenvalues of A and B), since A and B do not have common eigenvectors. See [21] for a detailed discussion of the argument and of its historical significance.

5.3.2 Context-Free Hidden Variables

More general hidden variable models have been rediscussed a few decades later, especially by and after J. Bell. Let me first present the class of models called (in the modern terminology) "context-free" or "noncontextual" hidden variables models. The idea is that one should consider only the correlations between results of measurements on a given system for *sets of commuting observables*. Indeed only such measurements can be performed independently and in any possible order (on a single realization of the system), and without changing the statistics of the outcomes. Any such given set of observables can be thought as a set of classical observables, but of course this classical picture is not consistent from one set to another.

Thus the idea is still that a hidden variable $\mathfrak{v}$ assigns to any observable A a definite outcome $a = f(A, \mathfrak{v})$ as in (5.34). This assumption is often called "value definiteness" (VD). To be compatible with quantum mechanics, the outcome a must be one of the eigenvalues of the operator A. To a given pure quantum state ψ corresponds a probability distribution $p_\psi(\mathfrak{v})$ over the probability space $\mathfrak{V}$. It is such that the quantum expectation value of any observable corresponds to the probabilistic expectation of the corresponding outcome over the hidden variable distribution

$$\langle \psi | A | \psi \rangle = \int_{\mathfrak{V}} p_\psi(\mathfrak{v}) \, f(A, \mathfrak{v}) \tag{5.36}$$

However the too strong constraint (5.35) should be replaced by the more realistic constraint, valid only for the set of outcomes $\{ f(A, \mathfrak{v}) \}$ for compatible observables (commuting operators)

$$\text{If } A_1 \text{ and } A_2 \text{ commute, then} \quad \left\{ \begin{array}{c} f(A_1 + A_2, \mathfrak{v}) = f(A_1, \mathfrak{v}) + f(A_2, \mathfrak{v}) \\ \text{and} \\ f(A_1 A_2, \mathfrak{v}) = f(A_1, \mathfrak{v}) f(A_2, \mathfrak{v}) \end{array} \right. \tag{5.37}$$

Moreover, these conditions are extended to any family $\mathcal{F} = \{ A_i, \ i = 1, 2, \cdots \}$ of commuting operators.

The models are context-free or non contextual. The term "context free" means that the outcome a for the measurement of the first observable A_1 is supposed to be independent of the choice of the other compatible observable A_2. In other word, the outcome of a measurement depends only of the choice of observable and of the state of the hidden variable, but not of the "context" of the measurement, that is of the other quantities measured at the same time.

Here I presented purely deterministic hidden variable models. We shall discuss the possibility that a is a random variable (with a probability law fixed by $\mathfrak{v}$) later.

5.3.3   Gleason's Theorem and Contextuality

5.3.3.1   Gleason's Theorem Excludes General Context-Free Models

These kind of models seem more realistic. However, they are immediately excluded by Gleason's theorem [50]. This was already pointed out and discussed by J. Bell in [14] (in fact before his famous theorem that I shall discuss later). Indeed, if to any $\mathfrak{v}$ is associated a function $f_\mathfrak{v}$, defined over the set of observables by

$$f_\mathfrak{v} \quad ; \qquad A \to f_\mathfrak{v}(A) = f(A, \mathfrak{v}) \tag{5.38}$$

which satisfy the consistency conditions (5.37), this condition is true for any family of commuting projectors $\{ P_i \}$, whose outcome in 0 or 1

$$P \text{ projector such that } P = P^\dagger = P^2 \quad \Longrightarrow \quad f_\mathfrak{v}(P) = 0 \text{ or } 1 \tag{5.39}$$

In particular, this is true for the family of projectors $\{P_i\}$ onto the vector of any orthonormal basis $\{\vec{e}_i\}$ of the Hilbert space $\mathcal{H}$ of the system. This means simply that defining the function f on the unit vectors $\vec{e}$ by

$$f(\vec{e}) = f_{\mathfrak{v}}(P_{\vec{e}}) \quad , \qquad P_{\vec{e}} = |\vec{e}\rangle\langle\vec{e}| \tag{5.40}$$

(remember that $\mathfrak{v}$ is considered fixed), this function must satisfy for any orthonormal basis

$$\{\vec{e}_i\} \quad \text{orthonormal basis} \qquad \Longrightarrow \qquad \sum_i f(\vec{e}_i) = 1 \tag{5.41}$$

while we have for any unit vector

$$f(\vec{e}) \;=\; 0 \text{ or } 1 \tag{5.42}$$

This contradicts strongly Gleason's theorem (see Sect. 4.4.2), as soon as the Hilbert space of the system $\mathcal{H}$ has dimension $\dim(\mathcal{H}) \geq 3$. Indeed, (5.41) means that the function f is a frame function (in the sense of Gleason), hence is continuous, while (5.42) (following from the fact that f is function on the projectors) means that f cannot be a continuous function. So

$$\dim(\mathcal{H}) \geq 3 \quad \Longrightarrow \quad \begin{array}{c} \text{no context-free deterministic hidden variable model} \\ \text{is compatible with quantum mechanics} \end{array} \tag{5.43}$$

5.3.3.2 The Special Case of $n = 2$

Gleason's theorem does not apply to the case $\dim(\mathcal{H}) \;=\; n \;=\; 2$. It is in fact quite easy to construct a context-free hidden variable model that describes all the observable of the 2-level system (the q-Bit). Here is a variant of a model by J. Bell. Any pure state $|\psi\rangle$ can be represented as a vector $\vec{n} \;=\; (\sin(\theta)\cos(\varphi), \sin(\theta)\sin(\varphi), \cos(\theta))$ of the unit 2-sphere in 3 dimensions through the Bloch sphere representation

$$|\vec{n}\rangle = \cos(\theta/2)|\uparrow\rangle + \sin(\theta/2)e^{i\varphi}|\downarrow\rangle \tag{5.44}$$

The algebra of observables is generated by the Pauli matrices and any self-adjoint physical operator A can be written as

$$A = \alpha\,\mathbf{1} + \vec{\beta}\cdot\vec{\sigma} \quad , \qquad \vec{\sigma} = (\sigma_x, \sigma_y, \sigma_z) \tag{5.45}$$

$\vec{\beta} \cdot \vec{\sigma}$ is the traceless part of the operator A, with $\vec{\beta} = (\beta_x, \beta_y, \beta_z)$ a real vector. One has

$$\langle \vec{n} | \vec{\beta} \cdot \vec{\sigma} | \vec{n} \rangle = \vec{\beta} \cdot \vec{n} \qquad (5.46)$$

so that the eigenvalues and eigenvectors of A are

$$a_\pm = \alpha \pm |\vec{\beta}| \quad , \qquad |\psi_\pm\rangle = |\vec{n}_\pm\rangle \quad , \qquad \vec{n}_\pm = \pm \vec{\beta}/|\vec{\beta}| \qquad (5.47)$$

One can take as hidden variables the unit vectors $\vec{v}$

$$\mathfrak{V} = S_2 \quad , \qquad \mathfrak{v} = \vec{v} \qquad |\vec{v}| = 1 \qquad (5.48)$$

with the outcomes

$$f(A, \vec{v}) = \begin{cases} \alpha + |\vec{\beta}| & \text{if } \vec{\beta} \cdot \vec{v} \geq 0, \\ \alpha - |\vec{\beta}| & \text{if } \vec{\beta} \cdot \vec{v} < 0. \end{cases} \qquad (5.49)$$

and for the probability distribution associated to the pure quantum state $|\vec{n}\rangle$ the distribution on the sphere with support on the

$$p_{\vec{n}}(\vec{v}) = \frac{1}{\pi} \begin{cases} \vec{n} \cdot \vec{v} & \text{if } \vec{n} \cdot \vec{v} \geq 0, \\ 0 & \text{if } \vec{n} \cdot \vec{v} < 0. \end{cases} \qquad (5.50)$$

5.3.3.3   Probabilistic Models = Quantum Mechanics

One may consider also partially deterministic hidden variable models, where to a hidden variable instance $\mathfrak{v}$ is attached a probability law $p(\ ; \mathfrak{v})$ defined as the probability that the outcome of a measurement of the observable A is a

$$\mathfrak{v} \to p(a|A; \mathfrak{v}) = \text{probability that measurement of } A \to a \qquad (5.51)$$

so that the expectation value of A with respect to the law $p(\ ; \mathfrak{v})$ and the quantum expectation value are

$$\mathbb{E}[A|\mathfrak{v}] = \int da \ a \ p(a|A; \mathfrak{v}) \quad , \qquad \langle \psi | A | \psi \rangle = \int_{\mathfrak{V}} p_\psi(\mathfrak{v}) \ \mathbb{E}[A|\mathfrak{v}] \qquad (5.52)$$

For such a model to be context-free, it must satisfy $\mathbb{E}[F(A)|\mathfrak{v}] = \int da \ F(a) \ p(a|A; \mathfrak{v})$. Then using Gleason's theorem again, it is easy to shown that the only consistent and minimal realization is to take for hidden variable the state vector itself, and for the probability law the law that is given by the Born rule itself (full quantum indeterminism)

$$\upsilon = \psi \quad , \qquad p(a|A; \psi) = \langle \psi | \delta(a\mathbf{1} - A) | \psi \rangle \qquad (5.53)$$

Therefore there is no need for any "sub-quantum" classical indeterminism.

Gleason's theorem is a very serious problem for the idea of hidden variables. It excludes the hypothesis that the values of all the possible observables of a quantum system are unknown to us but preexist the act of observation. Such a concept is (I think) often called the "strong realism hypothesis".

However, some remaining possibilities may still be considered, that correspond to a weaker notion of realism. The two main ones are

1. There are still context-free hidden variables, but they describe only some specific subset of the quantum correlations, not all of them.
2. There are hidden variables, but they are fully contextual.

I now discuss two famous cases where the first option has been explored, but appears to be still problematic. The second option raises also very serious questions, that will be shortly discussed in Sect. 5.4.

5.3.4 The Kochen-Specker Theorem

The first option is related to the idea that some subsets of the correlations of a quantum system have a special status, being related to some special explicit "elements of reality" (the "be-ables" or "maybe-ables" in the terminology of J. Bell), in contrast to the ordinary observables which are just "observ-ables". Thus a question is whether for a given quantum system there are some finite families of non commuting observables which can be associated to context-free hidden variables.

In fact the problems with non-contextual hidden variable models have been shown to arise already for very small such subsets of observables, first by S. Kochen and E. Specker [67] and by J. Bell in [14]. This is the content of the Kochen-Specker theorem. This theorem provides in fact examples of *finite families* of unit vectors $\mathcal{E} = \{\vec{e}_i\}$ in a Hilbert Space $\mathcal{H}$ (over $\mathbb{R}$ or $\mathbb{C}$) of finite dimension (dim($\mathcal{H}$) $= n$), such that it is impossible to find any frame function such that

$$f(\vec{e}_i) = 0 \text{ or } 1 \quad \text{and} \quad (\vec{e}_{i_1}, \cdots, \vec{e}_{i_n}) \text{ orthonormal basis} \implies \sum_{a=1}^{n} f(\vec{e}_{i_a}) = 1$$
$$(5.54)$$

The original example of [67] involves a set with 117 projectors in a three dimensional Hilbert space, that generates the group of symmetry of a 3d polyhedra and is a very nice example of non-trivial 3d geometry. Simpler examples in dimension

$n = 3$ and $n = 4$ with a smaller number of projectors have been provided by several authors (Mermin, Babello, Peres, Penrose). I refer to [71] for details and I shall not discuss more these examples and their significance. But all these examples show that the non–contextual character of quantum correlations is a fundamental feature of quantum mechanics.

5.3.5 The Bell-CHSH Inequalities and Local Realism

5.3.5.1 The Local Realism Hypothesis

Another important example, where the relations between contextuality and locality are discussed, is the situation of the so-called Einstein-Poldovski-Rosen (EPR) paradox. It was first considered by J. Bell in his famous 1964 paper [13]. Consider a bipartite system S that consists of two causally independent subsystems S_1 and S_2, for instance a pair of time-like separated photons in a Bell-like experiment. We are interested in the correlations between the result of independent measurements on S_1 and S_2. If A is some observable for the system S_1 and B some observable for the system S_2, this pair of observables is compatible, since A and B, (or more exactly $A \otimes 1$ and $1 \otimes B$) commute. Thus in quantum mechanics the result of a measurement on S_1 should depend on the state of the whole system S and of the choice of the observable A, but it should not depend on the measurement made on S_2. In other word, the result of a measurement on S_1 might depend on the local context of S_1 but it should not depend on the context of S_2.

Following J. Bell, let us assume that some hidden variables model (some "elements of reality") underly the bipartite system S, and that, in the spirit of the argument by EPR, this model is local in the sense that it is S_1-versus-S_2 context free. This means that the "sub-quantum" state of the whole system S is assumed to be described by some hidden variable $\mathfrak{v}$. The result a of the measurement of A on S_1 is determined (or obeys a probabilistic law) that depends on the hidden variable $\mathfrak{v}$, on the observable A chosen, and possibly of the local context of the measurement on S_1, but *not on the local context of the measurement done on* S_2. Similarly, the result b of the measurement of B on S_2 is depends on the hidden variable $\mathfrak{v}$, on the observable B chosen, and possibly of the local context for S_2, but *not on the local context for* S_1.

For a purely deterministic (dispersion free) hidden variable model with such a constraint of locality, this means that a hidden variable $\mathfrak{v}$ assigns determined outputs a and b to the measurements of A and B (respectively on S_1 and S_2)

$$\mathfrak{v} \quad \rightarrow \quad a = f_1(A, \mathfrak{v}) \,, \; b = f_2(B, \mathfrak{v}) \tag{5.55}$$

For a partially deterministic local hidden variable model, a hidden variable $\mathfrak{v}$ assigns *independent probability distributions* for the outcomes a and b of the measurements of A and B (respectively on $\mathcal{S}_1$ and $\mathcal{S}_2$)

$$\mathfrak{v} \quad \rightarrow \quad p_1(a|A;\mathfrak{v}) \, , \ p_2(b|B;\mathfrak{v}) \tag{5.56}$$

Since in general we have seen that on a single subsystem no context-free hidden variable model is compatible with quantum mechanics, in general A and B must be understood as

$$A \ = \ \text{measured observable} + \text{local context for } \mathcal{S}_1 \tag{5.57}$$

$$B \ = \ \text{measured observable} + \text{local context for } \mathcal{S}_2 \tag{5.58}$$

A notable exception (in fact the only one) is the situation where the subsystems $\mathcal{S}_1$ and $\mathcal{S}_2$ are 2-level quantum systems (qBits, spins 1/2, photons with two polarization states. In this case we may forget about the local contexts and take (5.55) at face value. They will not change the argument leading to the Bell-CHSH inequalities anyway.

In any case, hidden variables models for a bipartite system that assign outcome to separate measurements on the two subparts according to (5.55) or (5.56) are denoted usually "local hidden variable models". I tend to find this denomination a bit misleading, since for such models the "hidden variables" are in general still contextual. The denomination "local" indicates that the output of a measurement is assumed to depend on the hidden variable only through the local context of this measurement. This is the hypothesis of "local realism". One might perhaps rather call it the hypothesis of "local contextuality", since it defines "locally-contextual-only hidden variables models", or "hidden variables models that satisfy contextual local realism". Since the denominations "local realism" and "local hidden variable models" are standard I shall use them.

5.3.5.2 The Bell-CSHS Inequality

Let me now recall the derivation of the famous Bell-CHSH inequality. In a general local hidden variable model, a quantum state ψ of $\mathcal{S}$ corresponds to some probability distribution $q_\psi(\mathfrak{v})$ over the hidden variables $\mathfrak{v}$. $q_\psi(\mathfrak{v})$ represents our ignorance about the "elements of reality" of the system. If this description is correct, the probability for the pair of outcomes $(A, B) \rightarrow (a,b)$ in the state ψ is given by the famous representation

$$p_\psi(a,b|A, B) = \sum_{\mathfrak{v}} q_\psi(\mathfrak{v}) \, p_1(a|A;\mathfrak{v}) \, p_2(b|B;\mathfrak{v})) \tag{5.59}$$

with p_1 and p_2 the outcome probabilities as in (5.56) It is this peculiar form which implies the famous Bell and BHSH inequalities on the correlations between observables on the two causally independent subsystems. If we consider for observables for $\mathcal{S}_1$ (respectively $\mathcal{S}_2$) two (not necessarily commuting) projectors P_1 and P_1' (respectively Q_2 and Q_2'), with outcome 0 or 1, and take for observables

$$A = 2P_1 - 1 \quad A' = 2P_1' - 1 \quad B = 2Q_1 - 1 \quad B' = 2Q_1' - 1 \tag{5.60}$$

The outcomes a, a', b and b' are -1 or 1. Let us (the experimentalist) perform a series of experiments on an ensemble of independently prepared instances of the bipartite system $\mathcal{S}$, choosing randomly with equal probabilities $1/4$ to measure (A, B), (A', B), (A, B') or (A', B'), and combine the results to compute the average

$$\langle M \rangle = \langle AB \rangle - \langle AB' \rangle + \langle A'B \rangle + \langle A'B' \rangle \tag{5.61}$$

The same argument than the argument used in Sect. 5.2.2.1 when discussing the Tsirelson's bound, using the general inequality

$$a, a', b, b' \in [-1, 1] \quad \Longrightarrow \quad a(b - b') + a'(b + b') \in [-2, 2] \tag{5.62}$$

and the fact that for the hidden variable model the outcomes a, a', b and b' are a priori well defined for each instance of $\mathfrak{v}$ so that one can use (5.59) implies the Bell-CSHS inequality

$$-2 \leq \langle M \rangle \leq 2 \tag{5.63}$$

For a bipartite quantum mechanical state that consists of two q-Bits, this inequality is known to be violated for some simple quantum states, in particular the fully entangled single state $|\psi\rangle = (1/\sqrt{2})(|1\rangle \otimes |0\rangle - |0\rangle \otimes |1\rangle)$ and some adequate choice of observables. Indeed $\langle M \rangle$ is known to saturate the Tsirelon's bound $|\langle M \rangle| \leq 2\sqrt{2}$ for such states.

From the quantum mechanics point of view, the reason for the violation of the bound (5.63) is simple. Assuming that all quantum states give probabilities of the form expected from the general local hidden variable model (5.59) and that the subsystems probabilities $p_1(a|A; \mathfrak{v})$ and $p_2(b|B; \mathfrak{v})$ obey the quantum rules and are representable by density matrices for the subsystems (this is the maximal assumption allowed by local contextuality) means that any quantum state (mixed or pure) ω of the whole system could be represented by a density matrix of the form

$$\rho_\omega = \sum_\mathfrak{v} q_\omega(\mathfrak{v}) \, (\rho_1(\mathfrak{v}) \otimes \rho_2(\mathfrak{v})) \tag{5.64}$$

where $q_\omega(\mathfrak{v}) \geq 0$ is a density probability over $\mathfrak{V}$ and $\rho_1(\mathfrak{v})$ and $\rho_2(\mathfrak{v})$ density matrices relative to the subsystems $\mathcal{S}_1$ and $\mathcal{S}_2$. Such mixed states for a bipartite system are called separable states. But it is well known that not all mixed states of a bipartite systems are separable, in particular pure entangled states are not separable. In fact finding a criteria for characterizing separate states of a general bipartite or multipartite quantum system is a very interesting problem in quantum information.

I am not going to discuss the many and very interesting generalizations and variants of Bell inequalities (for instance the spectacular GHZ example for tripartite systems) and the possible consequences and tests of contextuality. I will not review either all the experimental tests of violations of Bell-like inequalities in various contexts, starting from the first experiments by Clauser, and those by Aspect et al., up to the most recent ones, that have basically closed all the possible loopholes in the hypothesis leading to the Bell-like inequalities. All the experimental results are in full agreement with the predictions of standard Quantum Mechanics and more precisely of Quantum Electro Dynamics. See for instance [71] for a recent and very complete review.   ·

5.3.6 Contextual Models

Let me discuss briefly the last possibility: relax completely the requirement of non-contextuality and consider fully contextual hidden variable models.

5.3.6.1 Bell's Simple Model

Full contextuality for a hidden variable model amounts to assume that the output a for the measurement of an observable A is determined by the full context of the measurement, i.e. all the other compatible measurements and operations that are performed independently on the system. Let me consider the simple case of a finite dimensional Hilbert space. One can reduce the situation to the case where all these operations are ideal projective measurements, and thus consider as "the context" a complete family[2] of compatible projectors $\mathcal{P} = \{P_i\}_{i=1,K}$ onto orthogonal subspaces $\mathcal{E}_i$ so that

$$P_i P_j = \delta_{ij} P_i \quad , \quad \mathbf{1} = \sum_i P_i \quad \text{i.e.} \quad \mathcal{H} = \bigoplus_i \mathcal{E}_i \quad , \quad \mathcal{E}_i \perp \mathcal{E}_j \qquad (5.65)$$

The projectors are not necessarily of rank 1 (projector onto pure states) so that $K \leq n = \dim(\mathcal{H})$.

$$\text{context} := \{P_j\}_{j=1,K} = \mathcal{P} \qquad (5.66)$$

---

[2]In some discussions the context may include additional parameters (not included into the hidden variables) like the details of the apparatus used, etc.

Then the output $a_i = 0, 1$ of a measurement of some $P_i \in \mathcal{P}$ is determined by the hidden variable $\mathfrak{v}$ through a function that depends on all the P_j's, not only on P_i.

$$a_i = f(P_i | \{P_j\}; \mathfrak{v}) = f_i(\mathcal{P}; \mathfrak{v}) \tag{5.67}$$

The only constraint is that

$$f_i(\mathcal{P}; \mathfrak{v}) = 0, 1 \quad , \qquad \sum_i f_i(\mathcal{P}; \mathfrak{v}) = 1 \tag{5.68}$$

and that, if a quantum pure state $|\psi\rangle$ corresponds to a hidden variable distribution p_ψ, one has

$$\langle \psi | P_i | \psi \rangle = \int_{\mathfrak{V}} p_\psi(\mathfrak{v}) \, f_i(\mathcal{P}; \mathfrak{v}) \tag{5.69}$$

The l.h.s. corresponds to the Born rule and should be independent on the context, i.e. of the choice of the others P_j's, $j \neq i$.

As noticed by J. Bell in [14], this is trivial to satisfy. Just take as hidden variable the quantum state vector itself, $|\psi\rangle$, plus a uniformly distributed real random variable $X \in [0, 1]$.

$$\mathfrak{v} = (|\psi\rangle, X) \tag{5.70}$$

From the probabilities $q_j = q_j(\mathcal{P}; |\psi\rangle) = \langle \psi | P_j | \psi \rangle$ of getting $a_j = 1$ in the state $|\psi\rangle$ (given by Born's rule), one can decompose the interval $[0, 1[$ into the successive disjoint intervals $\mathcal{I}_| = [X_{j-1}, X_j[$ with $X_j = \sum_{k=1,j} q_k$. The output functions are defined as

$$a_i = f_i(\mathcal{P}; \mathfrak{v}) = \begin{cases} 1 & \text{if } X \in \mathcal{I}_i = [X_{i-1}, X_i[, \\ 0 & \text{otherwise.} \end{cases} \tag{5.71}$$

and the probability distribution for a quantum state $|\phi\rangle$ is the Dirac distribution on the space of quantum states (the projective hyperplane in $\mathcal{H}$) times the uniform distribution over $[0, 1[$ for X

$$\int_{\mathfrak{V}} p_{|\phi\rangle}(\mathfrak{v}) = \int_{|\psi\rangle} \delta(|\psi\rangle, |\phi\rangle) \int_0^1 dX \tag{5.72}$$

Thus one reobtains easily the standard output quantum probabilities

$$p_i = \langle \phi | P_i | \phi \rangle \tag{5.73}$$

from the principle "Born rule in, Born rule out".

This model is very simple indeed, and was considered by J. Bell as too trivial and not physical. Indeed it amounts to standard quantum mechanics itself. The random variable X has no particular physical or "'ontological" status. It is just introduced to give an explicit representation of the quantum probabilities as probabilities on an abstract classical probability space. In a mathematical language, X implements the Kolmogorov extension theorem, which states that any abstract probabilistic process—provided it satisfies some obvious self-consistency conditions—can be represented on some explicit probability space Ω. But on the other hand this model is so generic (well, one has to extend it to infinite dimensional Hilbert spaces, see below) that one may suspect that any contextual hidden variable model might be reduced to this one, if one wants to keep it fully compatible with quantum mechanics.

5.3.6.2 de Broglie-Bohm Pilot-Wave Models

This is a more elaborate model which contains the dynamics of the Schrôodinger equation. It was first proposed by L. de Broglie, and rediscovered and developed by D. Bohm. Let me recall the model for the simple and standard example of a non relativistic particle in a scalar potential U. In position space and in the Schrödinger picture a pure quantum state $|\psi\rangle$ is given by the wave function $\psi = \psi(\vec{q}, t)$. The basic idea is that the continuity equation for the probability density ρ and the probability current $\vec{j}$

$$\frac{\partial \rho}{\partial t} + \vec{\nabla} \cdot \vec{j} = 0 \qquad (5.74)$$

with

$$\rho = |\psi|^2 \quad , \quad \vec{j} = \frac{i\hbar}{2m}(\psi \vec{\nabla} \bar{\psi} - \bar{\psi} \vec{\nabla} \psi) \qquad (5.75)$$

can be rewritten as the flow equation for the density distribution $\tilde{\rho} = \tilde{\rho}(\vec{q}, t)$ of particles driven by a vector field $\vec{V} = \vec{V}(\vec{q}, t)$ that derives from the wave function itself

$$\frac{\partial \tilde{\rho}}{\partial t} + \vec{\nabla}(\vec{V}\tilde{\rho}) \quad \text{with} \quad \vec{V} = \frac{\vec{j}}{\rho} = \frac{\hbar}{m}\vec{\nabla}(\arg(\psi)) = \frac{i\hbar}{2m}\left(\frac{\vec{\nabla}\bar{\psi}}{\bar{\psi}} - \frac{\vec{\nabla}\psi}{\psi}\right) \qquad (5.76)$$

Thus a hidden variable model consists in taking as sub-quantum degrees of freedom (hidden variable) the wave function itself $\psi = \psi(\vec{q}, t)$ (the "pilot wave") and the position vector $\vec{x} = \vec{x}(t)$ of a "real" particle, that will be the observable degree of freedom.

$$\mathfrak{v} = (\psi, \vec{x}) \tag{5.77}$$

The particle behaves as a passive scalar carried along the time dependent flow $\vec{V}$. The vector flow $\vec{V}$ is given by the gradient of the phase of the wave function ψ, that obeys the Schrödinger equation ($U = U(\vec{q})$ is the external scalar potential)

$$i\hbar \frac{\partial \psi}{\partial t} = -\frac{\hbar^2}{2m} \Delta \psi + U \psi \tag{5.78}$$

The dynamics for $\vec{x}$ is

$$\frac{d\vec{x}}{dt} = \vec{V}(\vec{x}, t) \tag{5.79}$$

and the initial probability distribution $\tilde{\rho}_\psi(\vec{x}, t = 0)$ for the particle in position space, that corresponds to the ensemble that describe a quantum state $|\psi\rangle$, has to be taken equal to the probability density function given by the Born rule

$$\tilde{\rho}_\psi(\vec{x}, t) = |\psi(\vec{x}, t)|^2 \qquad \text{at} \qquad t = 0 \tag{5.80}$$

It is easy to see that this relation (relating the probability distribution for the particle $\vec{x}$ to the pilot wave function ψ) stays valid at all times t, if the dynamics of the hidden variable $\mathfrak{v}$ is given by the Schrödinger equation (5.78) for ψ and the pilot wave equation (5.79) for $\vec{x}$.

With these notations for a contextual hidden variable theory, the de Broglie-Bohm theory describes the observables of position of the quantum particle. Thus the "context" is the set of projections over position eigenstates

$$\text{Context} = \mathcal{Q} = \{ P_{\vec{q}} = |\vec{q}\rangle\langle\vec{q}|; \ \vec{q} \in \mathbb{R}^d \} \tag{5.81}$$

The output rule is (the hidden variable being $\mathfrak{v} = (\psi, \vec{x})$)

$$f(P_{\vec{q}} | \mathcal{Q}; \mathfrak{v}) = \delta(\vec{x} - \vec{q}) \tag{5.82}$$

and the probability distribution associated to a quantum state $|\phi\rangle$ is

$$p_\phi(\mathfrak{v}) = p_\phi(\psi, \vec{x}) = \delta(\phi, \psi) |\psi(\vec{x})|^2 \tag{5.83}$$

One has indeed for any function A of the position operator $\vec{Q}$

$$\langle \phi | A(\vec{Q}) | \phi \rangle = \int_{\psi, \vec{x}} p_\phi(\psi, \vec{x}) A(\vec{x}) \tag{5.84}$$

The interesting property of the model is that its dynamics reproduces the average quantum dynamics. This means that the expectation value of the momentum operator P on a state $|\phi\rangle$ is given by the probabilistic average of the classical momentum $\vec{p} = m\dot{\vec{x}}$ of the "classical" particle

$$\langle \phi | \vec{P} | \phi \rangle = m \int_{\psi, \vec{x}} p_\phi(\psi, \vec{x}) \, \dot{\vec{x}} \tag{5.85}$$

The model can be trivially extended to more that one particle. Then the pilot wave function $\psi(\vec{x}_1, \vec{x}_2, \cdots)$ depends on the position vectors of all the particles, and takes into account the non-local correlations between the positions of the "classical" particles resulting from entanglement.

For a single particle in d=1 dimension, the de Broglie-Bohm model is in fact equivalent to the simple Bell's model discussed in the previous section. Indeed instead of the hidden variable x (the position of the particle), one can consider the variable X

$$X = \int_{-\infty}^{x} dy \, |\psi(y)|^2 \tag{5.86}$$

and switch to the hidden variables $\mathfrak{u}$

$$\mathfrak{v} = (\psi, x) \quad \rightarrow \quad \mathfrak{u} = (\psi, X) \tag{5.87}$$

For any quantum state the variable X is now uniformly distributed on the interval $[0, 1]$ and it has no dynamics

$$p_\phi(\psi, X) = \delta(\phi, \psi) \, 1 \quad , \qquad \dot{X} = 0 \tag{5.88}$$

5.3.6.3 Stochastic Models and Nelson Stochastic Mechanics

A question often discussed for these models is whether there is a good reason to take as initial distribution $\tilde{\rho}_\phi$ for the position $\vec{x}$ the quantum probability distribution $|\phi|^2$. Indeed, the evolution process for ψ and $\vec{x}$ and the evolution equation (5.76) are valid for $\tilde{\rho} \neq |\phi|^2$ as well.

One solution is to modify the evolution equation for $\vec{x}$ by adding some randomness, so that the general evolution equation for $\tilde{\rho}$ contains a diffusion term and a drift term, in such a way that any initial distribution $\tilde{\rho}$ relax irreversibly towards the quantum probability distribution $\tilde{\rho}_\phi = |\phi|^2$ at large time. The asymptotic equilibrium dynamics at large time, although non-deterministic, is physically indistinguishable for the deterministic model, as far as position observables are concerned.

In such models, the evolution equation for x becomes a stochastic equation of the form

$$dx = V\, dt + D\, dt + v\, dB_t \qquad (5.89)$$

where V is the driving term of the original model, D is a drift term, dB_t a Brownian process, and v a diffusion constant (it may depend on x). The drift term D is adjusted to the diffusion term so that the evolution equation takes the form (for v independent of x and t)

$$\frac{\partial \tilde{\rho}}{\partial t} + \vec{\nabla}(\vec{V}_\phi\, \tilde{\rho}) + v\, \rho_\phi\, \Delta(\tilde{\rho}/\rho_\phi) \qquad (5.90)$$

It is clear that the dynamics is now irreversible, and that it should relax towards the distribution where the diffusion term $\Delta(\tilde{\rho}/\rho_\phi)$ vanishes, namely the quantum probability distribution $\tilde{\rho} = \rho_\phi = |\phi|^2$.

These kind of models have been proposed by D. Bohm and his collaborators, who looked for physical models where the dynamics of the particles and the diffusion process was the result of some non-trivial microscopic sub-quantum dynamics, resulting in short time relaxation towards the equilibrium quantum distribution.

A particular case is the so-called *stochastic quantum mechanics* proposed by E. Nelson [81]. It amounts to a stochastic dynamics of the form (5.89), with a special choice for the diffusion coefficient

$$v = \frac{\hbar}{2m} \qquad (5.91)$$

This particular choice has some advantages. For a free particle (external potential $U = 0$), the dynamics of the particle is a Brownian process with some characteristics and statistics similar to those of the trajectories in the Feynman path integral formulation. The momenta of the particle $p = m\dot{q}$ becomes a stochastic variable with short time correlations of a white noise, with the right variance in order to reproduce as a statistical law the Heisenberg uncertainty relations, etc.

Other mechanisms that might produce relaxation of a general distribution function towards the quantum one have been suggested, based on the idea of coarse graining over microscopic sub-quantum degrees of freedom and of the notion of effective wave functions. See [100, 101] for a simple version, and [40] for another more detailed formulation and for details.

5.3.6.4   An Adiabatic Argument

Finally, let me indicate another argument (F. David, Unpublished, 2014) that justifies the choice of the quantum distribution measure $\rho_\phi = |\phi|^2$ as hidden variable distribution $\tilde{\rho}$. Let me start from a given quantum state ϕ_0, and to simplify take it to be an energy eigenstate of a time symmetric Hamiltonian H of the quantum system, so that the dynamics of the position hidden variable $\vec{x}$ is trivial ($\dot{\vec{x}} = 0$). Now let me perturbs adiabatically the dynamics along a cycle in the space of Hamiltonians, starting and ending at H

$$H \rightarrow H(t) = H + \epsilon\,\delta H(t\epsilon) \tag{5.92}$$

with $\delta H(0) = \delta H(1) = 0$, as done in the computation of the Berry phase. In the adiabatic limit $\epsilon \rightarrow$ after one cycle ($T = 1/\epsilon$) one ends in the same quantum state ϕ_0 (up to a phase, including the Berry phase, of course). But one can also study the adiabatic dynamics of the particle $\vec{x}(t)$ and argue that it is non trivial after one cycle, namely

$$\vec{x}(0) \rightarrow \vec{x}(T = 1/\epsilon) \neq \vec{x}(0) \tag{5.93}$$

The final position depends of the adiabatic cycle. Therefore, with the exception of the one dimensional case $d = 1$, adiabatic transformations mix the particle positions and the quantum distribution measure $\rho_\phi = |\phi|^2$ is the only probability distribution $\tilde{\rho}$ that is invariant under arbitrary adiabatic cyclic transformations, and hence natural under general unitary Hamiltonian evolutions.

5.3.6.5 The Problems with Contextual Models

Of course the main feature of these models is that they are contextual. They are appropriate, and indeed equivalent to standard quantum mechanics, as long as one wants to describe the properties of a quantum system that depend (at a given time) of a given family of compatible observables. In the pilot-wave models these observables (the context) are the position observables, and these models single out the space of positions of the particles (the configuration space) as playing a specific role.

The original pilot wave model manages to treat also (some aspects of) the single momentum operator P, but none of these contextual models will succeed in treating simultaneously, together with the position operators Q, Q^2, Q^3, etc. all the momentum operators P, P^2, P^3, etc. and the combinations of P's and Q's, in particular the energy operator (the Hamiltonian) H. This impossibility is ensured by the no-go theorems that we discussed (Gleason's theorem, Kochen-Specker's theorem, etc.).

It is indeed possible to construct a hidden variable model that deals with the P, P^2, P^3, etc. but it will be uncorrelated (and ontologically incompatible) with the model for the Q, Q^2, Q^3, etc. In order to construct a consistent hidden variable model, one must first know which physical quantities one is going to measure, and the description of the possible outputs of your measurements in terms of preexisting actual values for the hidden variables will depend on this choice! In the pilot wave model, the position x's variables can be considered as the subsets of hidden variables that you can actually observe, the remaining wave function ψ, i.e. the pilot wave, being actually the really hidden variable that you cannot observe. This description is valid (and sometime quite sensible and useful for physics or chemistry), but only in the context of position observables. To give a firm ontological status to these x variable (namely, stating that they are "element of reality" that preexists the measurements and determine the results of the measurement) is not really possible.

It is sometimes advocated that the position observables have a special status and should be considered as really more fundamental. But I think that this point of view is difficult to defend, especially in high energy physics and in quantum field theory, where the position $\vec{x}$ as well as the time t are not—and cannot be—physical observables!

Finally let me mention that this issue of contextuality and of hidden variables has been also discussed for quantum measurement processes. See for instance [41]. One has to associate also hidden variables to the measurement apparatus, in order to construct a hidden variable model for quantum measurements. However, in fact one has to introduce hidden variables common to the measurement system and the measurement apparatus, and to assume that there are preexisting variables that correlates the choice of the observed quantity by the observer and the results of the measurements. In particular, this means that the basic independence assumption in deriving the Bell inequalities—namely that the two observer can choose independently whatever they measure—is violated. The former—there are elements of reality that "exist" but always conspire to reproduce the results of quantum mechanics—is often denoted "superdeterminism". I shall not discuss more this point here, but for many physicists superdeterminism leads to formidable philosophical problems and to no predictive powers.

5.4 Summary Discussion on Quantum Correlations

Let me now try to summarize what can be learnt from the last two sections on quantum correlations and the notions of contextuality and non-locality.

The significance and the consequences of Bell's inequalities and their extensions, and of the Kochen-Specker-like theorem's for quantum theory have been enormously discussed, and some debates are still going on. It is not the purpose of these notes to review and summarize all these discussions that took place at the physics level and at the philosophical level. Let me just try to make some simple remarks.

5.4.1 Locality and Realism

As discussed in Sect. 5.3, the assumption of context-free value definiteness is clearly not tenable, from Gleason's theorem. This means that one must be very careful when discussing quantum physics about correlations between results of measurements. To quote a famous statement by Y. Peres: "Unperformed experiments have no results" [85].

Trying to assign some special ontological status to a (finite and in practice small) number of observables to avoid the consequence of the Kochen-Specker theorem may be envisioned, but raises other problems. For instance, if one wants to keep the main axioms of QM, and non-contextuality, by using a finite number of observables, one would expect the quantum logic formalism would lead to QM on a finite division ring (a Galois field), but it is known that this is not possible (see the discussion in Sect. 4.3.2.2). Note however that relaxing some basic physical assumptions like reversibility and unitarity has been considered for instance in [99].

It is also clear that non-local quantum correlations are present in non-separable quantum states, highlighted by the violations of Bell's and CHSH-like inequalities (and their numerous and interesting variants). They represent some of the most non-classical and counter-intuitive features of quantum physics. In connexion with the discussions of the "EPR-paradox", this non-local aspect of quantum physics has been often—and are still sometimes—presented as a contradiction between the principles of quantum mechanics and those of special relativity. This is of course not correct.

The concept of local realism, which is incompatible with quantum mechanics and known to be excluded by experiments, is indeed different from the concept of locality used in relativistic quantum field theory. As discussed in Sect. 5.3.5, local realism corresponds to the property of "local contextuality" for hidden variable models. This idea was advocated by Einstein and assumed to hold by Einstein, Poldovski and Rosen in their EPR 1935 paper. It means that two causally independent systems can be assigned separate and individual (but locally contextual) hidden variables ("elements of reality") and that classical correlations between these local hidden variables are sufficient to explain the quantum correlations between the two systems when they are entangled.

Locality in quantum field theory is something different, and corresponds to the concepts of local events (localized in space and time) and of the causal relations between these events as related to the geometry of space-time), in particular of there independence for space-separated points. This is the concept of locality and causality as formulated by A. Einstein in 1905 in the theory of special relativity, and then extended to general relativity. These requirements of causality and locality are necessary in quantum theory in order to formulate consistent relativistic quantum field theories, as briefly presented in Sect. 3.8. They are constraints on the observables (the operators) of the quantum theory, rather than constraints on the (tentative hidden variable description of the) quantum states. They imply that no information (causal effects) can propagate faster than light, and thus imply the "non-signaling" property of quantum information science. This is the reason why EPR-like experiments and the violations of Bell-like inequalities should not be considered as a signal of some "spooky-at-a-distance physical action" at work in quantum operations over entangled multipartite systems.

$$\text{locality in QFT} \neq \text{local realism}$$

In Sect. 5.3 I discussed also the status of fully contextual models, that would correspond to some form of "non-local contextual realism", and explained why they are in fact quite problematic. I shall come back shortly to these issues in Sect. 5.6, when discussing the relations between the formalisms and the interpretations.

5.4.2   Chance and Correlations

Now let me return to the quantum correlations discussed in Sect. 5.2. Contrary to classical physics, in quantum physics there is an irreducible quantum indeterminism and uncertainty in the description of a quantum system. Not all the physical observables can be characterized uniquely and independently at the same time. This essential feature reflects itself in the Heisenberg uncertainty principle, and can be formulated for instance as N. Bohr's "complementarity principle".

However, contrary to what could be expected, this does not mean that a quantum system is always more uncertain or "fuzzy" than a classical system. Indeed, the quantum correlations may very well be stronger than the corresponding classical correlations, when considering bipartites or multipartite systems. This is exemplified for instance by the quantum entropic inequalities (5.11) and (5.14) when compared to their classical analog, the entropic bound (5.12), and by the Tsirelson bound (5.26) compared to the B-CHSH inequality (5.63). Indeed, thanks to entanglement, quantum systems may be more correlated than what is expected classically when one assumes that correlations come only from classical, local and non-contextual "elements of reality" shared by the systems.

Nevertheless, the quantum correlations are still strongly controlled by the physical principles that we have discussed in the presentation of the formalisms, causality, reversibility and locality (in the causal sense), or by the information principle formulations shortly discussed in Sect. 5.1. The "super-strong" correlations (that can be build for instance using the Popescu-Rohrlich boxes discussed in Sect. 5.2.2.2) raise problems and it has not been possible to implement them in a physical theory. This can be represented by the little drawing of Fig. 5.6, where the set of quantum correlations (the red square) is shown to be larger than the set of classical correlations, but smaller that the set of all logically possible correlations.

This simple drawing illustrates why the theoretical works by J. Bell and its successors, besides their importance for our theoretical and philosophical understanding of what is and what is not quantum mechanics, turned out to have a significant and long term impact in science and technology. They played

Fig. 5.6 Schematic of the
worlds of classical
correlations, quantum
correlations and
"super-strong" unphysical
correlations

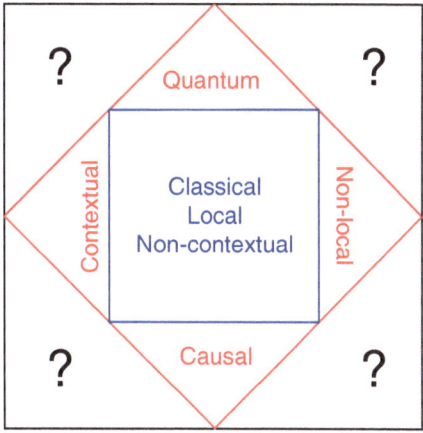

an important role in the rise of quantum information science, since they showed that, using quantum correlations and entanglement, it is possible to transmit and manipulate information, perform calculations and search processes, etc. in ways which are impossible by classical means, or which are much more efficient.

5.5 Measurements

5.5.1 What are the Questions?

Up to now I have not discussed much the question of quantum measurements. I simply took the standard point of view that (at least in principle) ideal projective measurements are feasible and one should look at the properties of the outcomes. The question is of course highly more complex. In this section I just recall some basic points about quantum measurements.

The meaning of the measurement operations is at the core of quantum physics. It was considered as such from the very beginning. See for instance the proceedings of the famous Solvay 1927 Congress [10], and the 1983 review by Wheeler and Zurek [109]. Many great minds have thought about the so called "measurement problem" and the domain has been revived in the last decades by the experimental progresses, which allows now to manipulate simple quantum system and implement effectively ideal measurements.

On one hand, quantum measurements represent one of the most puzzling features of quantum physics. They are non-deterministic processes (quantum mechanics predicts only probabilities for outcomes of measurements). They are irreversible processes (the phenomenon of the "wave-function collapse"). They reveal the irreducible uncertainty of quantum physics (the uncertainty relations). This makes quantum measurements very different from "ideal classical measurements".

On the other hand, quantum theory is the first physical theory that addresses seriously the problem of the interactions between the observed system (the object) and the measurement apparatus (the observer). Indeed in classical physics the observer is considered as a spectator, able to register the state of the real world (hence to have its own state modified by the observation), but without perturbing the observed system in any way. Quantum physics shows that this assumption is not tenable. Moreover, it seems to provide a logically satisfying answer[3] to the basic question: what are the minimal constraints put on the results of physical measurements by the basic physical principles.[4]

It is often stated that the main problem about quantum measurement is the problem of the uniqueness of the outcome. For instance, why do we observe a spin 1/2 (i.e. a q-bit) in the state $|\uparrow\rangle$ or in the state $|\downarrow\rangle$ when we start in a superposition $|\psi\rangle = \alpha|\uparrow\rangle + \beta|\downarrow\rangle$? However by definition a measurement is a process which gives one single classical outcome (out of several possible). Thus in my opinion the real questions, related to the question of the "projection postulate", are: (1) Why do repeated ideal measurements should give always the same answer? (2) Why is it not possible to "measure" the full quantum state $|\psi\rangle$ of a q-bit by a *single* measurement operation, but only its projection onto some reference frame axis?

Again, the discussion that follows is very sketchy and superficial. A good recent reference, both on the history of the "quantum measurement problem", a detailed study of explicit dynamical models for quantum measurements, and a complete bibliography, is the research and review article [4].

5.5.2 The von Neumann Paradigm

The general framework to discuss quantum measurements in the context of quantum theory is provided by J. von Neumann in his 1932 book [105, 106]. Let me present it on the simple example of the q-bit.

But before, let me insist already on the fact that this discussion will not provide a derivation of the principle of quantum mechanics (existence of projective measurements, probabilistic features and Born rule), but rather a self-consistency argument of compatibility between the axioms of QM about measurements and what QM predicts about measurement devices.

An ideal measurement involves the interaction between the quantum system S (here a q-bit) and a measurement apparatus M which is a *macroscopic object*. The idea is that M must be treated as a quantum object, like S. An ideal non destructive measurement on S that does not change the orthogonal states $|\uparrow\rangle$ and $|\downarrow\rangle$ of S (thus corresponding to a measurement of the spin along the z axis, S_z), correspond to introducing for a finite (short) time an interaction between S and M, and to start

---

[3]If not satisfying every minds, every times...

[4]Well... as long as gravity is not taken into account!

from a well chosen initial state $|I\rangle$ for $\mathcal{M}$. The interaction and the dynamics of $\mathcal{M}$ must be such that, if one starts from an initial separable state where $\mathcal{S}$ is in a superposition state

$$|\psi\rangle = \alpha\,|\uparrow\rangle + \beta\,|\downarrow\rangle \tag{5.94}$$

after the measurement (interaction) the whole system (object+apparatus) is in an entangled state

$$|\psi\rangle \otimes |I\rangle \qquad \rightarrow \qquad \alpha\,|\uparrow\rangle \otimes |F_+\rangle + \beta\,|\downarrow\rangle \otimes |F_-\rangle \tag{5.95}$$

The crucial point is that the final states $|F_+\rangle$ and $|F_-\rangle$ for $\mathcal{M}$ must be *orthogonal*[5]

$$\langle F_+|F_-\rangle = 0 \tag{5.96}$$

Of course this particular evolution (5.94) is unitary for any choice of $|\psi\rangle$, since it transforms a pure state into a pure state.

$$|\psi\rangle \otimes |I\rangle \qquad \rightarrow \qquad \alpha\,|\uparrow\rangle \otimes |F_+\rangle + \beta\,|\downarrow\rangle \otimes |F_-\rangle \tag{5.97}$$

One can argue that this is sufficient to show that the process has all the characteristic expected from an ideal measurement, within the quantum formalism itself. Indeed, using the Born rule, this is consistent with the fact that the state $\alpha|\uparrow\rangle$ is observed with probability $p_+ = |\alpha|^2$ and the state $\alpha|\downarrow\rangle$ with probability $p_- = |\beta|^2$. Indeed the reduced density matrices both for the system $\mathcal{S}$ and for the system $\mathcal{M}$ (projected onto the two pointer states) is that of a completely mixed state

$$\rho_S = \begin{pmatrix} p_+ & 0 \\ 0 & p_- \end{pmatrix} \tag{5.98}$$

For instance, as discussed in [105, 106], if one is in the situation where the observer $\mathcal{O}$ really observes the measurement apparatus $\mathcal{M}$, not the system $\mathcal{S}$ directly, the argument can be repeated as

$$|\psi\rangle \otimes |I\rangle \otimes |O\rangle \qquad \rightarrow \qquad \alpha\,|\uparrow\rangle \otimes |F_+\rangle \otimes |O_+\rangle + \beta\,|\downarrow\rangle \otimes |F_-\rangle \otimes |O_-\rangle \tag{5.99}$$

and it does not matter if one puts the fiducial separation between object and observer between $\mathcal{S}$ and $\mathcal{M} + \mathcal{O}$ or between $\mathcal{S} + \mathcal{M}$ and $\mathcal{O}$. This argument being repeated ad infinitum.

---

[5]as already pointed out in [105].

A related argument is that once a measurement has been performed, if we repeat it using for instance another copy $\mathcal{M}'$ of the measurement apparatus, after the second measurement we obtain

$$|\psi\rangle \otimes |I\rangle \otimes |I'\rangle \qquad \rightarrow \qquad \alpha \, |\uparrow\rangle \otimes |F_+\rangle \otimes |F_+'\rangle + \beta \, |\downarrow\rangle \otimes |F_-\rangle \otimes |F_-'\rangle \qquad (5.100)$$

so that we never observe both $|\uparrow\rangle$ and $|\downarrow\rangle$ in a successive series of measurements (hence the measurement is really a projective measurements). The arguments holds also if the outcome of the first measurement is stored on some classical memory device $\mathcal{D}$ and the measurement apparatus reinitialized to $|I\rangle$. This kind of argument can be found already in [80].

The discussion here is clearly outrageously oversimplified and very sketchy. For a precise discussion, one must distinguish among the degrees of freedom of the measurement apparatus $\mathcal{M}$ the (often still macroscopic) variables which really register the state of the observed system, the so called *pointer states*, from the other (numerous) microscopic degrees of freedom of $\mathcal{M}$, which are present anyway since $\mathcal{M}$ is a macroscopic object, and which are required both for ensuring decoherence (see next section) and to induce dissipation, so that the pointer states become stable and store in a efficient way the information about the result of the measurement. One must also take into account the coupling of the system S and of the measurement apparatus $\mathcal{M}$ to the environment $\mathcal{E}$.

5.5.3   Decoherence, Ergodicity and Mixing

As already emphasized, the crucial point is that starting from the same initial state $|I\rangle$, the possible final pointer states for the measurement apparatus, $|F_+\rangle$ and $|F_-\rangle$, are orthogonal. This is now a well defined dynamical problem, which can be studied using the theory of quantum dynamics for closed and open systems. The fact that $\mathcal{M}$ is macroscopic, i.e. that its Hilbert space of states in very big, is essential, and the crucial concept is *decoherence* (in a general sense).

The precise concept and denomination of quantum decoherence was introduced in the 1970s (especially by Zeh) and developed and popularized in the 1980s (see the reviews [66, 116]). But the basic idea seems much older and for our purpose one can probably go back to the end of the 1920 and to von Neumann's quantum ergodic theorem [104] (see [108] for the English translation and [51] for historical and physical perspective).

One starts from the simple geometrical remark [104] that if $|e_1\rangle$ and $|e_2\rangle$ are two *random unit vectors* in a N dimensional Hilbert space $\mathcal{H}$ (real or complex), their average "overlap" (squared scalar product) is of order

$$\overline{|\langle e_1 | e_2 \rangle|^2} \simeq \frac{1}{N} \quad , \qquad N = \dim(\mathcal{H}) \qquad (5.101)$$

hence it is very small, and for all practical purpose equal to 0, if N is very large. Remember that for a quantum system made out of M similar subsystems, $N \propto (N_0)^M$, N_0 being the number of accessible quantum states for each subsystem.

A simple idealized model to obtain a dynamics of the form (5.97) for $\mathcal{S}+\mathcal{M}$ is to assume that both $\mathcal{S}$ and $\mathcal{M}$ have no intrinsic dynamics and that the evolution during the interaction/measurement time interval is given by a interaction Hamiltonian (acting on the Hilbert space $\mathcal{H} = \mathcal{H}_\mathcal{S} \otimes \mathcal{H}_\mathcal{M}$ of $\mathcal{S} + \mathcal{M}$) of the form

$$H_{\text{int}} = |\uparrow\rangle\langle\uparrow| \otimes H_+ + |\downarrow\rangle\langle\downarrow| \otimes H_- \tag{5.102}$$

where H_+ and H_- are two *different* Hamiltonians (operators) acting on $\mathcal{H}_\mathcal{M}$. It is clear that if the interaction between $\mathcal{S}$ and $\mathcal{M}$ takes place during a finite time t, and is then switched off, the final state of the system is an entangled one of the form (5.97), with

$$|F_+\rangle = e^{\frac{t}{i\hbar}H_+}|I\rangle \quad , \qquad |F_-\rangle = e^{\frac{t}{i\hbar}H_-}|I\rangle \tag{5.103}$$

so that

$$\langle F_+|F_-\rangle = \langle I|e^{-\frac{t}{i\hbar}H_+} \cdot e^{\frac{t}{i\hbar}H_-}|I\rangle \tag{5.104}$$

It is quite easy to see that if H_+ and H_- are not (too much) correlated (in a sense that I do not make more precise), the final states $|F_+\rangle$ and $|F_-\rangle$ are quite uncorrelated with respect to each others and with the initial state $|I\rangle$ after a very short time, and may be considered as random states in $\mathcal{H}_\mathcal{M}$, so that

$$|\langle F_+|F_-\rangle|^2 \simeq \frac{1}{\dim(\mathcal{H}_\mathcal{M})} \lll 1 \tag{5.105}$$

so that for all practical purpose, we may assume that

$$\langle F_+|F_-\rangle = 0 \tag{5.106}$$

This is the basis of the general phenomenon of *decoherence*. The interaction between the observed system and the measurement apparatus has induced a decoherence between the states $|\uparrow\rangle$ and $|\downarrow\rangle$ of $\mathcal{S}$, but also a decoherence between the pointer states $|F_+\rangle$ and $|F_-\rangle$ of $\mathcal{M}$.

Moreover, the larger $\dim(\mathcal{H}_\mathcal{M})$, the smaller the "decoherence time" beyond which $\langle F_+|F_-\rangle \simeq 0$ is (and it is often in practice too small to be observable), and the larger (in practice infinitely larger) the "quantum Poincaré recurrence time" (where one might expect to get again $|\langle F_+|F_-\rangle| \simeq 1$) is.

Of course, as already mentioned, this is just the first step in the discussion of the dynamics of a quantum measurement. One has in particular to check and to explain how, and under which conditions, the pointer states are quantum microstates which correspond to macroscopic classical-like macrostates, which can be manipulated, observed, stored in an efficient way. At that stage, I just paraphrase J. von Neumann (in the famous Chap. VI "Der Meßprozeß" of [105])

Die weitere Frage (...) soll uns dagegen nicht beschäftigen.

Decoherence is a typical quantum phenomenon. It explains how, in most situations and systems, quantum correlations in small (or big) multipartite systems are "washed out" and disappear through the interaction of the system with other systems, with its environment or its microscopic internal degrees of freedom. Standard references on decoherence and the general problem of the quantum to classical transitions are [115] and [92].

However, the underlying mechanism for decoherence has a well know classical analog: it is the (quite generic) phenomenon of *ergodicity*, or more precisely the *mixing property* of classical dynamical systems. I refer to textbooks such as [5] and [74] for precise mathematical definitions, proofs and details. Again I give here an oversimplified presentation.

Let us consider a classical Hamiltonian system. One considers its dynamics on (a fixed energy slice $H = E$ of) the phase space Ω, assumed to have a finite volume $V = \mu(\Omega)$ normalized to $V = 1$, where μ is the Liouville measure. We denote T the volume preserving map $\Omega \to \Omega$ corresponding to the integration of the Hamiltonian flow during some reference time t_0. T^k is the iterated map (evolution during time $t = k t_0$). This discrete time dynamical mapping given by T is said to have the *weak mixing property* if for any two (measurable) subsets A and B of Ω one has

$$\lim_{n \to \infty} \frac{1}{n} \sum_{k=0}^{n-1} \mu(B \cap T^k A) = \mu(B)\mu(A) \qquad (5.107)$$

The (weak) mixing properties means (roughly speaking) that, if we take a random point a in phase space, its iterations $a_k = T^k a$ are at large time and "on the average" uniformly distributed on phase space, with a probability $\mu(B)/\mu(\Omega)$ to be contained inside any subset $B \in \Omega$. See Fig. 5.7

Weak mixing is one of the weakest form of "ergodicity" (in a loose sense, there is a precise mathematical concept of ergodicity).

Now in semiclassical quantization (for instance using Bohr-Sommerfeld quantization rules) if a classical system has M independent degrees of freedom (hence its classical phase space Ω has dimension $2M$), the "quantum element of phase space" $\delta\Omega$ has volume $\delta V = \mu(\delta\Omega) = h^M$ with $h = 2\pi\hbar$ the Planck's constant. If the phase space is compact with volume $\mu(\Omega) < \infty$ the number of "independent quantum states" accessible to the system is of order $N = \mu(\Omega)/\mu(\delta\Omega)$ and should

Fig. 5.7 Graphical representation of the mixing property (very crude)

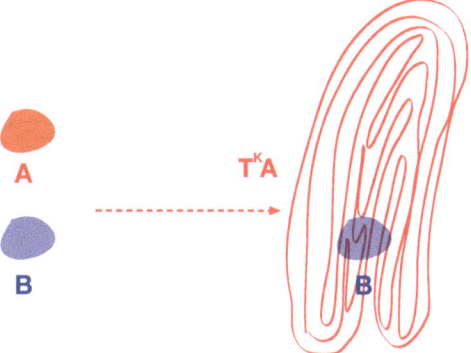

correspond to the dimension of the Hilbert space $N = \dim(\mathcal{H})$. In this crude semiclassical picture, if we consider two pure quantum states $|a\rangle$ and $|b\rangle$ and associate to them two minimal semiclassical subsets A and B of the semiclassical phase space Ω, of quantum volume δV, the semiclassical volume $\mu(A \cap B)$ corresponds to the overlap between the two quantum pure states through

$$\mu(A \cap B) \simeq \frac{1}{N}|\langle a|b\rangle|^2 \qquad (5.108)$$

More generally if we associate to any (non minimal) subset A of Ω a mixed state given by a quantum density matrix ρ_A we have the semiclassical correspondence

$$\frac{\mu(A \cap B)}{\mu(A)\mu(B)} \simeq N \operatorname{tr}(\rho_A \rho_B) \qquad (5.109)$$

With this semiclassical picture in mind (Warning! It does not work for all states, only for states which have a semiclassical interpretation! But pointer states usually do.) the measurement/interaction process discussed above has a simple semiclassical interpretation, illustrated on Fig. 5.8.

The big system $\mathcal{M}$ starts from an initial state $|I\rangle$ described by a semiclassical element I. If the system $\mathcal{S}$ is in the state $|\uparrow\rangle$, $\mathcal{M}$ evolves to a state $|F_+\rangle$ corresponding to F_+. If it is in the state $|\downarrow\rangle$, $\mathcal{M}$ evolves to a state $|F_-\rangle$ corresponding to F_-. For well chosen, but quite generic Hamiltonians H_+ and H_-, the dynamics is mixing, so that, while $\mu(F_+) = \mu(F_-) = 1/N$, typically one has $\mu(F_+ \cap F_-) = \mu(F_+)\mu(F_-) = 1/N^2 \ll 1/N$. Thus it is enough for the quantum dynamics generated by H_+ and H_- to have a quantum analog the classical property of mixing, which is quite generic, to "explain" why the two final states $|F_+\rangle$ and $|F_-\rangle$ are generically (almost) orthogonal.

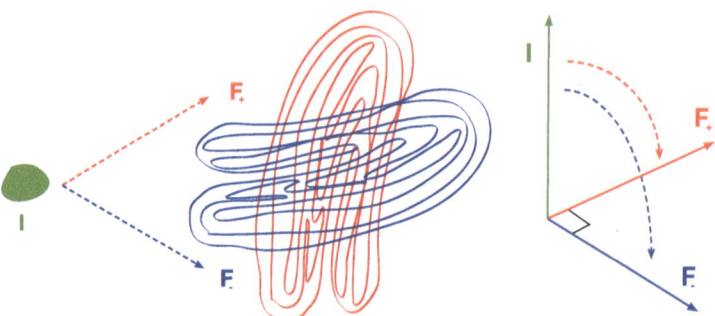

Fig. 5.8 Crude semiclassical and quantum pictures of the decoherence process (5.103)–(5.105)

5.5.4 Discussion

As already stated, the points that I tried to discuss in this section represent only a small subset of the questions about measurements in quantum mechanics. Again, I refer for instance to [4] and [71] (among many other reviews) for a serious discussion and bibliography.

I have not discussed more realistic measurement processes, in particular the so called "indirect measurements procedures", where the observations on the system are performed through successive interactions with other quantum systems (the probes) which are so devised as to perturb as less as possible the observed system, followed by stronger direct (in general destructive) measurements of the state of the probes. Another class of processes is the "weak measurements", which are a series of measurements that perturb weakly the measured system, combined with some adequate postselection process in order to define the "weak value" of an observable. Such measurement processes, as well as many interesting questions and experiments in quantum physics, quantum information sciences, etc. are described by the general formalism of POVM's (Positive Operator Valued Measure). I do not discuss these questions here.

In any case, important aspects of quantum measurements belong to the general class of problems of the emergence and the meaning of irreversibility out of reversible microscopic laws in physics (quantum as well as classical). See for instance [57].

The quantum formalism as it is presented in these lectures starts (amongst other things) from explicit assumptions on the properties of measurements. The best one can hope is to show that the quantum formalism is consistent: the characteristics and behavior of (highly idealized) physical measurement devices, constructed and operated according to the laws of quantum mechanics, should be consistent with the initials axioms.

One must be careful however, when trying to justify or rederive some of the axioms of quantum mechanics from the behavior of measurement apparatus and their interactions with the observed system and the rest of the world, not to make circular reasoning.

5.6   Formalisms, Interpretations and Alternatives to Quantum Mechanics

5.6.1   What About Interpretations?

In these notes I have been careful up to now not to discuss the interpretation issues of quantum mechanics. There are at least two reasons.

Firstly, I do not feel qualified enough to review and discuss all the interpretations of quantum mechanics that have been proposed and all the philosophical questions raised by quantum physics since its birth. This does not mean that I consider these question to be unimportant (nor that I never think about these).

Secondly, these lecture notes are focused on the presentation of the mathematical formalism of "standard quantum mechanics", and on the main approaches to present and justify physically the formalism. The point of view chosen is that of most physicists, chemists, mathematicians, computer scientists, engineers, etc who study or exploit the resources of the quantum world. This "operational point of view" consists in considering quantum mechanics as a theoretical framework that provides rules to compute the probabilities to obtain a given result/response when measuring/manipulating a quantum system as a function of its state. The concepts of "observables", "states" and "probabilities" being defined within the principles of the formalisms considered.

Of course this point of view may be already considered as a interpretational prejudice. So let me nevertheless come back to these issues of interpretations and try to make a few simple—and probably naïve—remarks.

As already explained in the introduction, I find the presentation and the discussion about the "interpretations of quantum mechanics" a bit confusing (even in some excellent reviews and textbooks) because they mix three different levels and topics: (1) the formalisations (mathematical formulations) of standard quantum theory, (2) the various interpretations of the formalisms and of their principles for quantum theory, (3) alternate theories, that try to obviate or solve in different ways some of the questions raised by quantum mechanics, but which are different physical theories of the quantum world, leading to different predictions than standard quantum mechanics in some regimes. These different concepts are usually presented at the same level, and denoted generically "interpretations of quantum mechanics", although in my opinion they may be quite different things.

5.6.2 Formalisms

Formulations such as the "canonical formalism", the path integral formalism, the algebraic and the quantum logic formalisms are in my opinion different mathematical representations of the same physical theory (or rather of the same physical framework in which one formulates the physical laws and physical models, in the same sense that classical physics is the framework for the physical laws of classical models). Their starting principles might be different, but they will lead to the same results or equivalent ones at the end. These different representations have their pros and cons, with different levels of mathematical rigor and of operability. However it is expected that they are not contradictory within their common domains of validity. Even if a real mathematician can claim—for instance—that no consistent quantum field theory has been rigorously constructed yet in four dimensional Minkowski space time, most physicists believe that this goal will be reached some day. Perhaps this will be done using the algebraic formalism, with some path integral and renormalization group. Perhaps some more powerful mathematical formalism will have to be used, quite different from the formalisms that we know at the moment.

5.6.3 Interpretations

Interpretations of quantum mechanics, such as the "Copenhagen interpretation" or the "many-worlds interpretation", are something slightly different. They do not challenge the present standard mathematical formulations of the theory, but rather insist on a particular point of view or a particular formulation of quantum mechanics as the best suited or the preferable one to consider and study quantum systems, and the quantum world, and they incorporate some more thoughts on the meaning of these principles. Thus in my opinion, rather than being formalizations of quantum mechanics, they should be considered as different and particular choices of points of view and of philosophical options to think about quantum mechanics and practice it. This does not mean that I am a conscious adept of post-modern relativism...

Let me first discuss the example of the so-called "Copenhagen Interpretation". Remember however that there is no clear cut definition of what "The Copenhagen Interpretation" is. Although they exist since the birth of quantum mechanics, the term "Copenhagen interpretation" was introduced only in 1955 by Heisenberg. I refer to the paper by Howard [62] for an historical and critical review of the history, uses and misuses of the concept. There are various points of view "from Copenhagen", some purely frequentists (quantum mechanics should apply to ensembles of systems only), some based on a more Bayesian concept and use of probabilities (quantum mechanics may be applied to single non-repeatable experiments on a single system). All this class of interpretations insist anyway on the fact that quantum mechanics deals only with the result of experiments and make predictions for the results of these experiments, not on some non-accessible underlying "reality". The presence of an "observer" (that does not need to be a conscious being) is thus important, either as a principle, or through theories of measurement processes or decoherence. Therefore I think the various Copenhagen interpretations may be considered as

Copenhagen = "quantum mechanics from a pragmatist point of view"

where "pragmatism" should be understood in the philosophical sense of pragmatism. Indeed these interpretations seems to be the most used (explicitly or implicitly) in experimental quantum physics and its applications.

At the other extreme one finds the "Many Worlds Interpretations". Again there are many variants, starting from the original proposal of Everett, its revised version by Wheeler and its presentation by B. Dewitt, who coined the term "many worlds" (see [37] for references), to the most recent views (see for instance [36]). The basic idea (as far as I understand) is to take seriously the concept of "the wave function of the universe" as the underlying concept of quantum mechanics, without reference to the observer, and to reinterpret or rederive the probabilistic features of quantum mechanics (in particular the Born rule and the projection postulate) as "relative" to some parts of this universal wave function, corresponding to some specific state of the observer for instance. From what I have seen, it seems that most proponents of the MW interpretations (but not all) agree on the fact that it does not allow to prove ab initio the Born law of probabilities. However, the MW point of view is often used in the field of cosmology and of quantum gravity and quantum cosmology, where the concept of "wave function of the universe" has to be tackled (for instance to discuss the Wheeler-DeWitt equation). It is also quite popular in some quantum information circles. I am not going to review or discuss more these interpretations, and refer to the recent proceedings+discussion book [91], which contains a modern presentation of the subject, and stimulating (and often contradictory) discussions and points of view by proponents and opponents. The role given to a (in practice unobservable) universal wave function that represents some "underlying reality" (but not in the sense of Einstein or Bell), while keeping the mathematical apparatus of quantum mechanics, is the central point for these interpretations. Therefore I think that most Many Worlds Interpretations may be considered as

Many Worlds = "quantum mechanics from a realist point of view"

Again, here realism is to be understood in its philosophical sense.

There is a whole spectrum of proposed interpretations that lie between these two main classes. Let me mention the "coherent history formulations" that insist on the fact that on should discuss and formulate quantum mechanics only in terms of "coherent histories", namely coherent semiclassical processes that admit some semiclassical description (the existence of such histories being justified by models of decoherence and coarse graining processes). I refer to [53] for details and references. This approach is also used in quantum gravity and quantum cosmology.

Another class of interpretations are the "modal interpretations" that try to make ontologically consistent the hidden variable theories based on pilot wave models by assigning a "modal status of reality" to the classical variables (such as the position of the particles driven by the wave function) so as to make them consistent with standard quantum mechanics. See [75] for a review of theses approaches.

There are many other (often overlapping) classes of interpretations that do not challenge in fact the mathematical formalism of quantum mechanics. A probably naïve and amateurish way to consider this "many-interpretations world of quantum mechanics" is to rather consider them as various point of view, from different philosophical perspectives, of the quantum formalism. Quantum mechanics is an impressive, beautiful and quite self consistent theoretical framework to describe physical phenomena, and it deserves several vistas to be appreciated and understood in its full majesty!

5.6.4   Alternatives

The interpretations that rely on the mathematical formulations of standard quantum mechanics should be clearly distinguished from another class of proposals to explain quantum physics that rely on modifications of the principles of quantum mechanics, and are thus different physical theories. These modified or alternative quantum theories deviate from "standard" quantum mechanics and should be experimentally falsifiable (and sometimes are already falsified).

This is the case of the various hidden-variables proposals, such as the de Broglie-Bohm pilot wave theories. It is often stated that the later formulations are equivalent to standard quantum mechanics, and fall into the previous category of interpretations. As discussed in Sect. 5.3.6, this is correct only in the restricted context of position observables (for the simple case of non-relativistic particles), and generally when dealing with a specified family of compatible observables, since these hidden variable models are contextual. However giving an ontological status to the position variable x of the particle driven by the pilot wave ψ, by considering x as an element of reality that can be related to what is observed, changes in my opinion the status of this model. It is now a different theoretical model than quantum mechanics, since some observables of quantum mechanics, like the impulsion p, are not described by this model, while the velocity of the particle $\dot{x}$ is a different variable that cannot be directly described by quantum mechanics.

This is also the case for the class of models known as "collapse models". See [48, 49] for the first models. In these models the quantum dynamics is modified by non-linear terms so that the evolution of the wave functions is not unitary any more (while the probabilities are conserved of course), and the "collapse of the wave function" becomes now a dynamical phenomenon. These models are somehow phenomenological, since the origin of these non-linear dynamical effects is quite ad hoc, they may be for instance justified by some quantum gravity effects. They predict a breakdown of the law of quantum mechanics for the evolution of quantum coherences and for decoherence phenomenon at large times, large distances, or in particular for big quantum systems (for instance large molecules or atomic clusters). Hence they can in principle be tested experimentally. At the present day, despite the impressive experimental progresses in the control of quantum coherences, quantum measurements, study of decoherence phenomenon, manipulation of information in quantum systems, no such violations of the predictions of standard QM and of unitary dynamics have been observed.

5.7 What About Gravity?

Another really important issue that I do not discuss in these lecture notes is quantum gravity. Again just a few simple remarks.

It is clear that the principles of quantum mechanics are challenged by the question of quantizing gravity. The challenges are not only technical. General relativity (GR) is indeed a non-renormalizable theory, and from that point of view a first and natural idea is to consider it as an effective low energy theory. After all, history tells us that in the development of nuclear and particle physics there has been several times (in the 1930, the 1940, the 1960...) theoretical false alarms and clashes between experiments and theory. Each time this led many great minds to question the principles of quantum mechanics themselves. However further development and understandings of the standard formalism (and experiments of course) allowed to solve these problems, so that quantum mechanics came out unscathed and even stronger. Since the 1970s and the construction of the standard model of strong and electroweak interactions the principles of quantum mechanics are not challenged any more.

However with gravity the situation is different. For instance the discovery of the Bekenstein-Hawking entropy of black holes, of the Hawking radiation, and of the "information paradox" shows that fundamental questions remain to be understood about the relation between quantum mechanics and the GR concepts of space and time. Indeed even the most advanced quantum theories available, quantum field theories such as non-abelian gauge theories the standard model, its supersymmetric and/or grand unified extensions, still rely on the special relativity concept of space-time, or to some extend to the dynamical but still classical concept of curved space-time of GR. It is clear that a quantum theory of space time will deeply modify, and even abolish, the classical concept of space-time as we are used to. Let me stress two important points.

Firstly, the presently most advanced attempts to build a quantum theory incorporating gravity, namely string theory and its modern extensions, as well as the alternative approaches to build a quantum theory of space-time such as loop quantum gravity (LQG) and spin-foam models (SF), rely mostly on the quantum formalism as we know it, but change the fundamental degrees of freedom (drastically and quite widely for string theories, in a more conservative way for LQG/SF). The fact that string theories offers some serious hints of solutions of the information paradox, and some explicit solutions and ideas, like holography and AdS/CFT dualities, for viewing space-time as emergent, is a very encouraging fact.

Secondly, in the two formalisms presented here, the algebraic formalism and the quantum logic formulations, it should be noted that space and time (as continuous entities) play a secondary role with respect to the concept of causality and locality/separability. I hope this is clear in the way I choose to present the algebraic formalism in Sect. 3 and quantum logic in Sect. 4. Of course space-time is essential for constructing physical theories out of the formalism. But the fact that it is the causal relations and the causal independence between physical measurement operations that are essential for the formulation of the theory is in my opinion a very encouraging signal that quantum theory may go beyond the classical concept of space-time.

Nevertheless, it may very well happens that (for instance) the information paradox is not solved by a sensible quantum theory of gravity, or that the concepts of causality and separability have to be rejected. Indeed one might imagine that no repeatable measurements are possible in a quantum theory of gravity, or that space being an "emergent" concept two separate sub-ensembles-of-degrees-of-freedom may never be considered as really causally independent. Then one might expect that the basic principles of quantum mechanics will not survive (and, according to the common lore, should be replaced by something even more bizarre, if not inexplicable...).

Bibliography

1. S. Adler, *Quaternionic Quantum Mechanics and Quantum Fields* (Oxford University Press, New York, 1995)
2. Y. Aharonov, D. Bohm, Significance of electromagnetic potentials in quantum theory. Phys. Rev. **115**, 485–491 (1959)
3. Y. Aharonov, P.G. Bergmann, J.L. Lebowitz, Time symmetry in the quantum process of measurement. Phys. Rev. **134**(6B), 1410–1416 (1964)
4. A.E. Allahverdyan, R. Balian, T.M. Nieuwenhuizen, Understanding quantum measurement from the solution of dynamical models (2012)
5. V.I. Arnold, A. Avez, *Ergodic Problems in Classical Mechanics* (Benjamin, New York, 1968)
6. V. Arnold, K. Vogtmann, A. Weinstein, *Mathematical Methods of Classical Mechanics.* Graduate Texts in Mathematics, 2nd revised edition (Springer, New York, 1989)
7. E. Artin, *Geometric Algebra*, vol. 3 (Interscience Publishers, New York, 1957)
8. G. Auletta, *Foundations and Interpretation of Quantum Mechanics* (World Scientific, Singapore, 2001)
9. G. Auletta, M. Fortunato, G. Parisi, *Quantum Mechanics* (Cambridge University Press, Cambridge, 2009)
10. G. Bacciagaluppi, A. Valentini, *Quantum Theory at the Crossroads, Reconsidering the 1927 Solvay Conference* (Cambridge University Press, Cambridge, 2012), arXiv:quant-ph/0609184
11. R. Baer, *Linear Algebra and Projective Geometry* (Dover, Mineola, 2005)
12. M. Bauer, Probabilité et processus stochastiques, pour les physiciens (et les curieux) (2009), http://ipht.cea.fr/Docspht/search/article.php?id=t09/324
13. J.S. Bell, On the Einstein-Podolsky-Rosen paradox. Physics **1**, 195 (1964)
14. J.S. Bell, On the problem of hidden variables in quantum mechanics. Rev. Mod. Phys. **38**, 447–452 (1966)
15. E.G. Beltrametti, G. Cassinelli, *The Logic of Quantum Mechanics.* Encyclopedia of Mathematics and Its Applications, vol. 15 (Addison-Wesley, Reading, 1981)
16. G. Birkhoff, J. von Neumann, The logic of quantum mechanics. Ann. Math. **37**, 823–843 (1936)
17. N.N. Bogoliubov, A.A. Logunov, A.I. Oksak, I.T. Todorov, *General Principles of Quantum Field Theory.* Number 10 in Mathematical Physics and Applied Mathematics (Kluwer Academic, Dordrecht, 1990)
18. G. Brassard, H. Buhrman, N.Linden, A.A. Méthot, A. Tapp, F. Unger, Limit on nonlocality in any world in which communication complexity is not trivial. Phys. Rev. Lett. **96**, 250401 (2006)

© Springer International Publishing Switzerland 2015   149
F. David, *The Formalisms of Quantum Mechanics*, Lecture Notes in Physics 893,
DOI 10.1007/978-3-319-10539-0

19. O. Bratteli, D.W. Robinson, *Operator Algebras and Quantum Statistical Mechanics*, vols. I and II (Springer, Berlin, 2002)
20. C. Brukner, Questioning the rules of the game. Physics **4**, 55 (2011)
21. J. Bub, Von Neumann's "no hidden variables" proof: a re-appraisal. Found. Phys. **40**, 1333–1340 (2010)
22. G. Chiribella, G.M. D'Ariano, P. Perinotti, Probabilistic theories with purification. Phys. Rev. A **81**, 062348 (2010)
23. G. Chiribella, G.M. D'Ariano, P. Perinotti, Informational derivation of quantum theory. Phys. Rev. A **84**, 012311 (2011)
24. B.S. Cirel'son, Quantum generalizations of Bell's inequality. Lett. Math. Phys. **4**, 93–100 (1980). doi:10.1007/BF00417500
25. J.F. Clauser, M.A. Horne, A. Shimony, R.A. Holt, Proposed experiment to test local hidden-variable theories. Phys. Rev. Lett. **23**, 880–884 (1969)
26. B. Coecke, Quantum picturalism. Contemp. Phys. **51**(1), 59–83 (2010)
27. C. Cohen-Tannoudji, B. Diu, F. Laloë. *Quantum Mechanics*, vols. 1 and 2 (Wiley, New York, 2006)
28. A. Connes, *Noncommutative Geometry* (Academic, New York, 1994)
29. A. Connes, M. Marcolli, *Noncommutative Geometry, Quantum Fields and Motives*, vol. 55 (AMS Colloquium Publications, AMS and Hindustan Book Agency, 2008)
30. R.T. Cox, Probability, frequency and reasonable expectation. Am. J. Phys. **14**(1), 1–13 (1946)
31. E. Crull, G. Bacciagaluppi, Translation of: W. Heisenberg, "Ist eine deterministische Ergänzung der Quantenmechanik möglich?" (2011). http://philsci-archive.pitt.edu/8590/
32. F. David, Importance of reversibility in the quantum formalism. Phys. Rev. Lett. **107**, 180401 (2011)
33. F. David, A short introduction to the quantum formalism[s] (2012). Saclay preprint t2/042, arXiv:1211.5627
34. B. de Finetti, *Theory of Probability* (Wiley, New York, 1974)
35. P. de la Harpe, V. Jones, An introduction to C*-algebras: Chapters 1 to 9, vol. 1 (1995). Publications internes de la Section de mathématiques de l'Université de Genève
36. D. Deutsch, *The Fabric of Reality: The Science of Parallel Universes– and Its Implications* (Allen Lane, New York, 1997)
37. B.S. DeWitt, R.N. Graham (eds.), *The Many-Worlds Interpretation of Quantum Mechanics*. Princeton Series in Physics (Princeton University Press, Princeton, 1973)
38. P.A.M. Dirac, *The Principles of Quantum Mechanics*. International Series of Monographs on Physics (Clarendon, Oxford, 1930)
39. J. Dixmier, *Les algèbres d'opérateurs dans l'espace Hilbertien: algèbres de Von Neumann*, volume fasc. 25 of *Cahiers scientifiques*, 2e édition revue et augmenteé edition (Gauthier-Villars, Paris, 1969)
40. D. Dürr, S. Goldstein, N. Zanghì. Quantum equilibrium and the origin of absolute uncertainty. J. Stat. Phys. **67**, 843–907 (1992)
41. D. Dürr, S. Goldstein, N. Zanghì. Quantum equilibrium and the role of operators as observables in quantum theory. J. Stat. Phys. **116**, 959–1055 (2004)
42. W. Ehrenberg, R.E. Siday, The refractive index in electron optics and the principles of dynamics. Proc. Phys. Soc. B **62**, 8–21 (1949)
43. W Feller, *An Introduction to Probability Theory and Its Applications* (Wiley, New York, 1968)
44. C.A. Fuchs, Quantum foundations in the light of quantum information (2001)
45. C.A. Fuchs, Quantum mechanics as quantum information (and only a little more) (2002), arXiv:quant-ph/0205039v1
46. G. Gasinelli, P. Lahti, A theorem of Solér, the theory of symmetry and quantum mechanics. Int. J. Geom. Methods Mod. Phys. **9**(2), 1260005 (2012)
47. I.M. Gelfand, M.A. Naimark, On the embedding of normed rings into the ring of operators in Hilbert space. Mat. Sb. **12**, 197–213 (1943)

48. G.C. Ghirardi, A. Rimini, T. Weber, *A Model for a Unified Quantum Description of Macroscopic and Microscopic Systems* (Springer, Berlin, 1985)
49. G.C. Ghirardi, A. Rimini, T. Weber, Unified dynamics for microscopic and macroscopic systems. Phys. Rev. **D34**, 470 (1986)
50. A.M. Gleason, Measures on the closed subspaces of a Hilbert space. Indiana Univ. Math. J. **6**, 885–893 (1957)
51. S. Goldstein, J.L. Lebowitz, C. Mastrodano, R. Tumulka, N. Zanghi, Normal typicality and von Neumann's quantum ergodic theorem. Proc. R. Soc. A **466**, 3203–3224 (2010)
52. K.R. Goodearl, *Notes on Real and Complex C*-Algebras*. Shiva Mathematics Series (Shiva Publishing Ltd., Nantwitch, UK, 1982)
53. R.B. Griffiths, *Consistent Quantum Theory* (Cambridge University Press, Cambridge, 2002)
54. D. Gross, M. Müller, R. Colbeck, O.C.O. Dahlsten, All reversible dynamics in maximally nonlocal theories are trivial. Phys. Rev. Lett. **104**, 080402 (2010)
55. R. Haag, *Local Quantum Physics: Fields, Particles, Algebras* (Springer, Berlin, 1996)
56. R. Haag, D. Kastler, An algebraic approach to quantum field theory. J. Math. Phys. **5**, 848–861 (1964)
57. J.J. Halliwell, J. Pérez-Mercader, W.H. Zurek (eds.), *Physical Origins of Time Asymmetry* (Cambridge University Press, Cambridge, 1996)
58. L. Hardy, Quantum theory from five reasonable axioms (2001). Perimeter Institute preprint, arXiv:quant-ph/0101012
59. L. Hardy, Reformulating and reconstructing quantum theory (2011). Perimeter Institute preprint, arXiv:1104.2066
60. A.R. Hibbs R. P. Feynman, *Quantum Mechanics and Path Integral* (Dover, New York, 2010) [Emended edition by d. f. styer edition]
61. S.S. Holland, Orthomodularity in infinite dimensions; a theorem of M. Solèr. Bull. Am. Math. Soc. (N.S.) **32**, 205–234 (1995)
62. D. Howard, Who invented the "Copenhagen Interpretation"? A study in mythology. Philos. Sci. **71**(5), 669–682 (2004)
63. L. Ingelstam, Real Banach algebras. Arkiv för Mat. **5**, 239–270 (1964)
64. J.M. Jauch, *Foundations of Quantum Mechanics* (Addison-Wesley, Reading, 1968)
65. E.T. Jaynes, *Probability Theory: The Logic of Science* (Cambridge University Press, Cambridge, 2003)
66. E. Joos, H.D. Zeh, C. Kiefer, D. Giulini, J. Kupsch, I.-O. Stamatescu, *Decoherence and the Appearance of a Classical World in Quantum Theory*, 2nd edn. (Springer, Berlin, 2003)
67. S. Kochen, E.P. Specker, The problem of hidden variables in quantum mechanics. J. Math. Mech. **17**, 59–87 (1967)
68. A.N. Kolmogorov, *Foundations of Probability* (Chelsea Publishing Company, New York, USA, 1950)
69. F. Laloe, Do we really understand quantum mechanics? Am. J. Phys. **69**, 655 (2001)
70. F. Laloë, *Comprenons-nous vraiment la mécanique quantique?* (EDP Sciences, Paris, 2011)
71. F. Laloë, *Do We Really Understand Quantum Mechanics?* (Cambridge University Press, Cambridge, 2012)
72. L.D. Landau, E.M. Lifshitz, *Mechanics*, 3rd edn. (Butterworth-Heinemann, Oxford UK, Burlington MA, USA, 1976)
73. M. Le Bellac, *Quantum Physics* (Cambridge University Press, Cambridge, 2011)
74. A.J. Lichtenberg, M.A. Lieberman, *Regular and Chaotic Dynamics*. Applied Mathematical Sciences (Springer, Berlin, 1992)
75. O. Lombardi, D. Dieks, Modal interpretations of quantum mechanics, in *The Stanford Encyclopedia of Philosophy*, ed. by E.N. Zalta, spring 2014 edition (2014), http://plato.stanford.edu/entries/qm-modal/
76. G. Ludwig, *Foundations of Quantum Mechanics* (Springer, New York, 1985)
77. G.W. Mackey, *Mathematical Foundation of Quantum Mechanics* (Benjamin, New York, 1963)

78. F. Maeda, S. Maeda, *Theory of Symmetric Lattices*. Grundlehren der mathematischen Wissenschaften (Springer, Berlin, 1971)
79. L. Masanes, M. Mueller, A derivation of quantum theory from physical requirements. New J. Phys. **13**, 063001 (2011)
80. N.F. Mott, The wave mechanics of α-ray tracks. Proc. R. Soc. Lond. A **129**, 79–84 (1929)
81. E. Nelson, Derivation of the Schrödinger equation from Newtonian mechanics. Phys. Rev. **150**, 079–1085 (1966)
82. M.A. Nielsen, I.L. Chuang, *Quantum Computation and Quantum Information*, 10th anniversary edition (Cambridge University Press, Cambridge, 2010)
83. C.N. Parkinson, Parkinson's Law. The Economist, 19 November 1955
84. M. Pawlowski, T. Paterek, D. Kaszlikowski, V. Scarani, A. Winter, M. Zukowski, Information causality as a physical principle. Nature **461**, 1101–1104 (2009)
85. A. Peres, Unperformed experiments have no results. Am. J. Phys. **46**(7), 745 (1978)
86. A. Peres, *Quantum Theory: Concepts and Methods* (Springer, Berlin, 1995)
87. C. Piron, "Axiomatique" quantique. Helv. Phys. Acta **37**, 439–468 (1964)
88. C. Piron, *Foundations of Quantum Physics* (Benjamin, New York, 1976)
89. S. Popescu, D. Rohrlich, Quantum nonlocality as an axiom. Found. Phys. **24**, 379–385 (1994). doi:10.1007/BF02058098
90. S. Sakai, *C*-Algebras and W*-Algebras* (Springer, Berlin, 1971)
91. B. Saunders, J. Barrett, A. Kent, D. Wallace (eds.), *Many worlds? Everett, Quantum Theory and Reality* (Oxford University Press, Oxford, 2010)
92. M.A. Schlosshauer, *Decoherence and the Quantum-to-Classical Transition* (Springer, Berlin, 2007)
93. I. Segal, Irreducible representations of operator algebras. Bull. Am. Math. Soc. **53**, 73–88 (1947)
94. I. Segal, Postulates for general quantum mechanics. Ann. Math. **48**, 930–948 (1947)
95. M.P. Solèr, Characterization of Hilbert spaces with orthomodular spaces. Commun. Algebra **23**, 219–243 (1995)
96. R.F. Streater, *Lost Causes in and Beyond Physics* (Springer, Berlin, 2007)
97. R.F. Streater, A.S. Wightman, *PCT, Spin and Statistics and All That*. Landmarks in Mathematics and Physics (Princeton University Press, Princeton, 2000)
98. E.C.G. Stueckelberg, Quantum theory in real Hilbert space. Helv. Phys. Acta **33**, 727–752 (1960)
99. G. 't Hooft, A mathematical theory for deterministic quantum mechanics. J. Phys. Conf. Ser. **67**(1), 012015 (2007)
100. A. Valentini, Signal-locality, uncertainty, and the sub-quantum h-theorem, I. Phys. Lett. A **156**(5) (1991)
101. A. Valentini, Signal-locality, uncertainty, and the sub-quantum h-theorem, II. Phys. Lett. A **158**(1) (1991)
102. W. van Dam, Implausible consequences of superstrong nonlocality (2005). arXiv:quant-ph/0501159. Natural Computing, March 2013, vol. 12, issue 1, pp. 9–12
103. V.S. Varadarajan, *The Geometry of Quantum Mechanics* (Springer, New York, 1985)
104. J. von Neumann, Beweis des Ergodensatzes und des H-theorems in der neuen Mechanik. Z. Phys. **57**, 30–70 (1929)
105. J. von Neumann, *Mathematische Grundlagen der Quantenmechanik*. Grundlehren der mathematischen Wissenschaften, volume Bd. 38. (Springer, Berlin, 1932)
106. J. von Neumann, *Mathematical Foundations of Quantum Mechanics*. Investigations in Physics, vol. 2. (Princeton University Press, Princeton, 1955)
107. J. von Neumann, *Continuous Geometry*. Princeton Mathematical Series, vol. 25. (Princeton University Press, Princeton, 1960)
108. J. von Neumann, Proof of the ergodic theorem and the H-theorem in quantum mechanics. Eur. Phys. J. H **35**, 201–237 (2010). doi:10.1140/epjh/e2010-00008-5
109. J.A. Wheeler, W. Zurek, *Quantum Theory and Measurement*. Princeton Series in Physics (Princeton University Press, Princeton, 1983)

110. S. Weinberg, *The Quantum Theory of Fields, Volume 1: Foundations* (Cambridge University Press, Cambridge, 2005)
111. W.K. Wooters, in *Complexity, Entropy, and the Physics of Information*, ed. by W.H. Zurek (Addison-Wesley, Reading, 1990), pp. 39–46
112. A. Zee, *Quantum Field Theory in a Nutshell* (Princeton University Press, Princeton, 2003)
113. J. Zinn-Justin, *Quantum Field Theory and Critical Phenomena*. International Series of Monographs on Physics (Clarendon, Oxford, 2002)
114. J. Zinn-Justin, *Path Integrals in Quantum Mechanics*. Oxford Graduate Texts, pbk. ed edition (Oxford University Press, Oxford, 2010)
115. W.H. Zurek (ed.), Complexity, entropy, and the physics of information, in *Santa Fe Institute Studies in the Sciences of Complexity*, vol. 8 (Addison-Wesley, Redwood City, 1990)
116. W.H. Zurek, Decoherence and the transition from quantum to classical – revisited (2003), arXiv:quant-ph/0306072v1

Index

© Springer International Publishing Switzerland 2015
F. David, *The Formalisms of Quantum Mechanics*, Lecture Notes in Physics 893,
DOI 10.1007/978-3-319-10539-0